FUN FOR ALL

Carmen Mangiron, Pilar Orero & Minako O'Hagan (eds)

FUN FOR ALL

Translation and Accessibility Practices in Video Games

PETER LANG
Bern · Berlin · Bruxelles · Frankfurt am Main · New York · Oxford · Wien

Bibliographic information published by die Deutsche Nationalbibliothek
Die Deutsche Nationalbibliothek lists this publication in the Deutsche Nationalbibliografie; detailed bibliographic data is available on the Internet at ‹http://dnb.d-nb.de›.

British Library Cataloguing-in-Publication Data: A catalogue record for this book is available from The British Library, Great Britain

Library of Congress Cataloging-in-Publication Data

Fun for all : translation and accessibility practices in video games / Carmen Mangiron, Pilar Orero & Minako O'Hagan (eds.).
pages cm
ISBN 978-3-0343-1450-3
1. Multimedia systems–Research. 2. Audio-visual equipment–Technological innovations.
3. Translating and interpreting–Technological innovations. 4. Computer games–Programming.
5. Video games–Design. 6. Video games industry–Technological innovations.
I. Mangiron, Carmen. II. Orero, Pilar. III. O'Hagan, Minako, 1959-
P306.93.F86 2014
794.8'1526–dc23

2013049684

Cover design: Thomas Jaberg, Peter Lang AG

ISBN 978-3-0343-1450-3 br. ISBN 978-3-0351-0667-1 eBook

Hochfeldstrasse 32, CH-3012 Bern, Switzerland
info@peterlang.com, www.peterlang.com

Acknowledgements

First and foremost we would like to acknowledge and thank the contributors for sharing their work with us. Special mention must also be made to Hugh Tanton and Estel·la Oncins for their revising and formatting assistance.

We are also extremely grateful to the Master in Audiovisual Translation at the Universitat Autònoma de Barcelona (pagines.uab.cat/mtav/), offered by the Departament de Traducció i d'Interpretació (Department of Translation and Interpreting), for its financial help with the publication of this book. Needless to say, any errors and misinterpretations remain our sole responsibility.

The publication of this book has also been made possible thanks to a grant from the Spanish Ministry of Finance and Competivity no. FFI2012-39056-C02-01 Subtitling for the deaf and hard of hearing and audio description: new formats, and by the Catalan Government funds 2009SGR700.

Table of contents

Carmen Mangiron, Pilar Orero & Minako O'Hagan

Fun for All: Translation and Accessibility Practices in Video Games

Video games have evolved to become a pervasive format which is beyond entertainment, enjoyed by a broad group of people rather than as a niche activity by hardcore gamers. More recently casual and social games have been further pushing such a trend, turning digital games into a universal phenomenon. Recent game statistics in developed countries indicate the ratio of people in the population who play digital games surpassing well over the 50% mark. For example, the US-based Entertainment Software Association reported that 58% of Americans play video games in 2013 (ESA 2013). This suggests mainstreaming of video games and their potentially significant influence on societies as a whole, providing further impetus to research.

This edited volume arose from the first International Conference on Translation and Accessibility in Video Games and Virtual Worlds held at the Universitat Autònoma de Barcelona (UAB) on 2–3 December 2010, organised by the TransMedia Catalonia Research Group of UAB, which is part of the Centre for Ambient Intelligence and Accessibility in Catalunya (CAIAC). The focus of the conference was unique in its comprehensive approach combining game accessibility, game translation and localization and also related issues in broader environments of virtual worlds. Such combination was due to a broad view of accessibility, which not only encompasses users with functional diversity, but also those who due to age or skill are not able to play a game successfully, those who do not speak the original language of the game, and those who due to socioeconomic conditions cannot access games or virtual platforms. In addition, both game localization and accessibility are issues that ideally need to be taken into account early in the development process of a game, in order prevent costly modifications at a later stage. If accessibility and localization are considered since the conception of the game, it should not be too costly or cumbersome to translate the game into other languages or include accessibility features, such as the possibility to remap the controls or have intralingual subtitles.

Gathering together researchers in academia as well as practitioners from the industry, the conference provided an ideal forum to exchange different perspectives to add to the emerging areas of research in translation and media accessibility of digital content. The conference demonstrated how research on game localization and research on game accessibility are gathering momentum, albeit they are still relatively unexplored academic fields. To date, most of the research conducted on game accessibility has been carried out by academics from engineering and computer studies backgrounds, such as the Universally Accessible Games (UA-Games) Research Group at the Human-Computer Interaction Laboratory of ICS-FORTH in Greece, who have developed a number of universally accessible games, and a handful of industry professionals interested in the topic, such as the Game Accessibility Special Interest Group (SIG) at the International Game Developer Association (IGDA). Most information and guidelines on game accessibility can be found on specialised blogs and websites, and are addressed to developers, to help them make their games accessible to the widest range of population. However, there seems to be a lack of awareness by mainstream developers and publishers about accessibility issues and due to the pressures and time limits associated to the release of a game, accessibility features usually are not a priority. It could be argued that by not taking into account accessibility, developers are leaving behind an important segment of the population: people with disabilities, who according to the UN account for 15% of the word population (The World Bank, 2013). People with sensorial (visual, auditory), cognitive, and physical disabilities also can enjoy playing video games, as they are fun and provide hours of entertainment.

As regards the topic of game translation, we wish to share a few observations on the development of the domain that is beginning to emerge as a legitimate research field in Translation Studies. Following the official declaration of year one of game studies in 2001 (Aarseth, 2001), academic research on video games has multiplied, helped by its increasingly ubiquitous presence in modern life and, most of all, by its inherent richness, inspiring many a research avenues. When *The Game Localization Handbook* (Chandler, 2005) was first published, the domain was a somewhat esoteric practice, little known beyond the immediate circle of practitioners and gamers involved in fan translation of games. Despite its close link to software localization, which has been integrated into translator training at many universities for some time now, localization practices applied in video games have remained under-reported until recently in academia. Similarly, the localization industry has focussed mainly on business-oriented productivity software. For example, in the early days games were not even included in the official statistics of the software industry as they were categorised as consumer goods and part of the entertainment sector (Berry, 2008: 66).

After Game Localization found early difficulties gaining recognition both in industry and academia, it is now an essential part of the game industry which relies on global sales while rapidly establishing itself as an interdisciplinary area of research in academia. Contemporary games are sophisticated technological as well as cultural artefacts with multiple perspectives that lend themselves to study and analysis from many and distant fields: from Psychology to Neuroscience or Sociology. In Translation Studies game localization is often introduced as a hybrid mode of Audiovisual Translation (AVT) and Software Localization (Munday, 2008: 190) in an attempt to locate this newly emerged practice within an established category. The research trajectory so far in game localization illustrates how the field tagged itself to these pre-existing domains, motivated by the fact that video games are indeed audiovisual media and also that they are pieces of software. In the sense that both Audiovisual Translation and Software Localization are dynamic fields sensitive to advances on technology, the boundaries between the two currently separate divisions are increasingly becoming blurred. Game Localization is an illustrative example of such a practice encroaching into both territories.

Rather similar to the initial passage of audiovisual research, studies on Game Localization are developing from a largely descriptive nature to a deeper conceptualisation both by practitioners reflecting on their work and researchers applying theoretical frameworks borrowed and adapted from Translation Studies. So far the work conducted in Game Localization has highlighted game-specific features in relation to other types of translation work. For example, they may be regulatory and operational constraints under which this practice is customarily performed, including age ratings as well as censorship implications and the severe limitation on the availability of contextual information, as is the case for Dubbing, Subtitling and Voice-over practices where translators have partial access to the visual text or they translate preproduction material (Orero, 2005).

This practise is also shared in game translation for games released under a sim-ship (simultaneous shipment) model, where localised games are released together with the original game. In a sim-ship model, translators work with a product under development. In translation research contexts, three key facets of modern games, i.e. technological, socio-cultural, and its didactic dimensions, are starting to attract scholarly attention. Research may be directed towards implications of new types of user interface, sophisticated use of cinematics or audio channel, and the use of games to learn languages to name but a few. Similarly, socio-cultural issues are increasingly addressed by researchers focusing on aspects of gamer culture, such as fan translation, and also a wide range of cultural transformation required beyond the verbal dimension during the localization process. We believe that a more systematic and interdisciplinary approach bringing together academics from different disciplines with various research backgrounds and methodologies, such as Translation Studies, Media Studies, Psychology, Usability, Engineering and Computing, is required to promote further meaningful advances in both game localization and accessibility and also to avoid duplication of efforts and ideas.

The collection of papers we have assembled in this volume focuses on game translation and accessibility and includes some of those presented in the I International Conference on Translation and Accessibility in Video Games and Virtual Worlds together with fresh contributions received from elsewhere that are relevant to the conference theme. Because of the emerging nature of the topic and the academia-industry mix of contributors, we have applied a strict academic convention to research-based contributions selection, while deciding to also include valuable articles which had the origin in professional practices and industry-based observations. The emerging nature of the field is such that contributions by reflective practitioners are a complement to theoretically-supported arguments in academic papers, and the book aims at bridging the gap between academia and industry, theory and practice. In presenting this volume we aim to provide a snapshot of varied research interest on the topic in an attempt to further establish game localization research in Translation Studies while building on the increasingly diverse media accessibility research methodologies, formats and platforms.

The volume is divided in two sections: the first section includes four contributions on Game Accessibility and the second nine contributions on different issues affecting game translation and localization. The opening paper in the accessibility section is From Game Accessibility to Universally Accessible Games by Dimitris Grammenos. His contribution introduces the concept of universally accessible games as an approach to creating games which are designed to be concurrently accessible by people with a wide range of requirements and (dis)abilities. Universally accessible games can adapt their interface, gameplay and content, so that they can best serve the requirements of each gamer, under specific gaming conditions. Grammenos also provides an overview of a design methodology for creating universally accessible games, and introduces the concept of parallel game universes as a means of supporting multiplayer sessions amongst people with diverse (dis)abilities, so that players share the same game and are fully aware of each other while at the same time experiencing the game in a way that is optimally adapted to their needs.

The second contribution Translating Fun for All: Promoting Accessibility in Video Games, by Alberto Fernández Costales, highlights the important role Translation Studies can play in promoting Game Accessibility. Fernández Costales underlines the importance of a broader concept of accessibility aimed at promoting e-inclusion and allowing access to video games to the widest possible audience, including those who speak a different language and those who have special needs. He argues that the relationship between translation and accessibility in the context of the gaming industry needs to be further explored, and supports the idea that translation is a key element to fostering e-inclusion.

The third contribution in the accessibility section, Accessible games and education: Accessibility experiences with <e-Adventure>, is by Javier Torrente, Ángel del Blanco, Pablo Moreno-Ger, Iván Martínez-Ortiz, and Baltasar Fernández-Manjón, members of the <e-UCM> research group at the Department of Software Engineering and Artificial Intelligence at the Complutense University of Madrid, Spain. Torrente et al. engage with the topic of accessible educational games and explore the use of game creation tools with built-in accessibility features, an efficient way of including accessibility features that does not compromise development costs. In their article they argue that accessibility for educational games is a must, as this type of games must be inclusive and available to everyone, regardless of their capabilities, and present *eAdventure*, a game-authoring platform designed to facilitate the creation of educational point-and-click adventure games with built-in accessibility features. They also describe a pilot case study which consisted of adding the accessibility features available at the *eAdventure* platform to an existing game, *1492*, an educational game about Spanish history, which was tested by a blind player and a player with reduced mobility.

The fourth and last contribution in the accessibility section is by professional developer and accessibility expert Javier Mairena, author of the Spanish blog Videojuegos Accesibles, which provides information, news, and guidelines regarding Game Accessibility. Mairena outlines a set of recommendations for developing accessible video games, and lists the different accessibility features any game should include. He also provides a classification of the main disability user groups, the needs of whom should be taken into consideration when developing a game, and proposes solutions for different accessibility challenges.

The second section of the book, focusing on game translation and localization, opens with Ornella Lepre's contribution Divided by Language, United by Gameplay: An Example of Ludological Game Localization. Lepre tackles the issue of cultural adaptation and remakes from Japanese games for the international markets. She proposes a ludological approach to game localization with the objective of reproducing the gameplay experience of the original for the target players. This may imply changing the graphics, the setting or the soundtrack, among others. Taking the Japanese rhythm game *Osu! Tatakae! Ouendan!* (2005) and its localised version for the North American territory, *Elite Beat Agents* (2006) as a case study, Lepre concludes that two games can provide the same experience for players as long as there are no changes in gameplay, even if the language, the setting, the soundtrack, the cultural references, or even the story are modified.

The second paper, Translation Strategies and Video Game Translation, by Annelies Van Oers presents the case study of the Dutch localised version of the video game *Beyond Good and Evil* (2003). Van Oers's analysis focuses on the translation strategies applied by localisers to diegetic text – text that is part of the fictional world of the video game – and concludes that the most used strategy is, unexpectedly, literal translation, possibly due to the fact that literal translation takes less time and is sometimes the safest option in game localization when there is no context and the deadlines are tight.

The third paper, Translating the Onscreen Text Blindfolded: Possibilities and Impossibilities, by Gianna Tarquini, deals with the challenges posed by the common practice of having to translate a video game without having access to it. After defining game localization as a form of constrained translation, Tarquini presents a case study conducted on a relatively large database of outsourced video game translation projects, mainly from the English-Italian and English-French linguistic combinations. The author highlights the main constraints found when translating onscreen text and analyses the recurrent translation patterns applied. Tarquini concludes highlighting the need for further descriptive studies in game localization, which are crucial for supporting theoretical assumptions concerning game translation theory and are also necessary to prepare students for real-life professional practice.

The fourth paper in this section, Video Games and Fan Translations: A case study of *Chrono Trigger*, by Rafael Müller, explores the world of video game fan translation. Müller analyses the different Brazilian fan translations of the Japanese game *Chrono Trigger* (1995), which are based on the first official English translation carried out for Nintendo by Ted Woolsey in 1995. Müller also compares these translations with another fan translation by Anglophone fans, as well as the updated official English version, released in 2008. Müller's analysis focuses on the following aspects: dialogue additions and omissions; the re-creation of play on words; the renaming of characters and terminology; censored items; the deliberate use of regional expressions, and the modification of a character's speech style. He concludes highlighting the need for further studies exploring the interesting and growing phenomenon of game fan translations.

As its title indicates, the paper Terminology Management in Video Game Localization, by Xiaochun Zhang, tackles terminology management in game localization, an issue that has been to date overseen. Zhang argues that terminology management can be extremely beneficial for the game localization industry and discusses the processes and methods of managing terms in industrial practice. She concludes with an assessment of the benefits of establishing an online video game public terminology database.

Stephen Mandiberg's paper, Games, Localization, and Diaspora, approaches game localization from a Cultural Studies perspective. Mandiberg is critical about the way in which localization practices to date have failed to make games accessible to alternate populaces, such as communities in diaspora. After defining the concept of diaspora, he focuses on the example of the Chinese diaspora and argues that games localized for the Chinese locale are inaccessible to people from across the Chinese Diaspora, living in Hong Kong, Taiwan, Australia, or the United States. He proposes "omnilingual language implementation" in game localization as a solution to this problem, which would allow players to choose different language and dialect combinations or download language packs as required. Mandiberg also explores how crowdsourcing could contribute to solve this issue and facilitate the development of diasporic games.

The following contribution, by Oliver Carreira and Eugenia Arrés, Video Game Localization Training on Offer in Spanish Universities at an Undergraduate Level, claims that game localization training currently on offer at Spanish universities is insufficient, as most undergraduate degrees in Translation and Interpreting do not include this subject in their curricula, despite the growing industry need for trained localisers. After outlining a number of prerequisites for working in the game localization industry, Carreira and Arrés subsequently analyze the training currently on offer in Spanish universities. They also present the opinion of several localization professionals they interviewed about the training currently offered. They conclude proposing different strategies for introducing game localization into undergraduate curricula in Spain, in the hope that these strategies may contribute to bridging the gap between the Spanish educational framework and the employment market.

The last two contributions in the game translation section are by two experienced industry professionals working in the field of interactive media and game localization. Víctor Alonso's paper, New Challenges in Interactive Media Localization Projects, aims at providing guidance on both successful and unsuccessful methods for approaching game localization projects in a new global environment which is also locally focused. Alonso's paper attempts to explain the environment of constant change faced by game localization due to the challenge placed by the continuous innovation of the game industry. Alonso concludes that project management is an essential and strategic competency for companies and individuals working on localization, as it allows them to adapt and adjust to the ever evolving environment of game localization.

Finally, the last contribution in the volume, by Kate Edwards, with the title of Beyond Localization: An Overview of Game Culturalization, highlights the importance of the process of cultural adaptation or culturalization in game localization in order to avoid any negative backlash for game developers and publishers and to produce games with more locally-relevant content. After discussing the key role culture plays in game design, Edwards presents the different levels of game culturalization that can be applied in game localization and describes the different types of culturalization challenges, namely history, religion, ethnicity, and geopolitics. She concludes stating that culturalization should be an integral part of game localization, as it can ensure that the work and the creative vision of game developers can be enjoyed by as many cultures as possible.

The different contributions in this pioneering volume address the emerging fields of Game Accessibility and Game Localization from different angles, providing insightful information about these relatively unexplored academic areas with such close tights to the industry. However, many other aspects remain yet to be explored. Further research in Game Accessibility and Game Localisation can be beneficial for all, as it can help increase the potential market size for developers and publishers. Improving access to games can also foster the inclusion of different type of users, from language learners to the aged, including those with sensorial, motor or mental diversity and those who do not speak the original language of the game. The industry and academia should work hand in hand to keep advancing towards a society where video games and the interactive entertainment they provide are available to all, regardless of their abilities and the language they speak. A world where video games can truly provide fun for all.

Carmen Mangiron, Pilar Orero and Minako O'Hagan

Conventions used in this book

When a videogame is mentioned, the year of its first release is included in brackets. In the case of game franchises, a hyphen is used after the year to indicate that the series is ongoing.

References

Aarseth, Espen. 2001. 'Computer Game Studies, Year One'. Game Studies1, no. 1. Online at: <http://www.gamestudies.org/0101/editorial.html> (consulted 03.05.2012).

Berry, David M. 2008. *Copy, rip, burn: The politics of copyleft and open source*. London: Pluto Press.

Chandler, Heather M. 2005. *The Game Localization Handbook*. Massachusetts: Charles River Media.

Entertainment Software Association (ESA). 2013. 'Industry Facts'. Online at: <http://www.theesa.com/facts/index.asp> (consulted 20.09. 2013).

Munday, Jeremy. 2008. *Introducing Translation Studies: Theories and Applications* (2nd ed.). London and New York: Routledge.

Orero, Pilar. 2005. 'La traducción de entrevistas para voice-over' in Zabalbeascoa, Patrick, Laura Santamaría and Frederic Chaume (eds). *La traducción audiovisual. Investigación, enseñanza y profesión*. Granada: Comares: 213–222.

The World Bank. 2013. 'Disability: Overview'. Online at: <http://web.worldbank.org/WBSITE/EXTERNAL/TOPICS/EXTSOCIALPROTECTION/EXTDISABILITY/0,,contentMDK:21151218~menuPK:282706~pagePK:210058~piPK:210062~theSitePK:282699,00.html> (consulted 20.09.2013).

Section I: Game accessibility

DIMITRIS GRAMMENOS

From Game Accessibility to Universally Accessible Games

Computer games are usually quite demanding in terms of the motor, sensor and mental skills needed for interaction control, and they often require the mastering of inflexible, relatively complicated input devices and techniques. These factors often render computer games inaccessible to a large percentage of people including, but not limited to, people with special needs. In this context, this paper introduces the concept of 'Universally Accessible Games' as a novel approach to creating games which are proactively designed to be concurrently accessible by people with a wide range of diverse requirements and (dis)abilities. The global vision is that through Universally Accessible Games, all people will be able to play videogames and have fun together, collaborating or competing on an equal basis, irrespective of their individual characteristics and preferences, and the technologies used. The distinctive characteristic of a Universally Accessible Game is that it can adapt its interface, gameplay and content to best serve the requirements of each gamer, under specific gaming conditions. This paper also provides an overview of a design methodology for creating Universally Accessible Games and suggests the concept of 'Parallel Game Universes' as a means of supporting multiplayer sessions amongst people with diverse (dis)abilities, where players share the same game and are fully aware of each other while at the same time experiencing the game in an optimally adapted way.

1. Introduction

Traditionally, the term "accessibility" has been associated with people with physical, sensory or mental disabilities. Throughout this paper, the term is also used to encompass people with "diversified needs" due to: (a) the environment in which they operate; (b) the devices and software they use or (c) their particular personal skills and/or preferences. In this context, the term "(video) game accessibility" is used to describe a situation in which a person is able to play a game even whilst having "diversified needs", or whilst playing under "limiting conditions" which may include permanent and temporary physical, sensory and mental disabilities. Some indicative examples of *diversified needs* include playing a game:

- in a foreign language
- left-handed, or whilst being able to use just one hand
- in a very bright / dark or loud / quiet environment
- whilst on the move
- as a novice, casual, tired, very young or very old player
- using "alternative" devices (e.g. using a touchpad instead of a mouse or a virtual keyboard)
- using devices with limitations or problems (e.g. with a broken keyboard key, on a mobile phone screen or on a TV screen located too far away)

Hopefully by this point it should have become clear that accessibility does not only address "physical, sensory and mental disabilities". It is a much broader term, which is used to refer to "providing to *ANY* person access to *SOMETHING*". In other words, if accessibility was a door key, then it would be a skeleton *("pass-partout")* key that could open *ANY* door, not just doors with a specific type of lock. In fact, what is most of the time considered a 'disability' is usually just a matter of statistics or 'environmental variables'.

For example, 70–90% of the world's population are right-handed. In practice, this means that objects and tools created by any one of them are more likely to better fit their needs than the needs of left-handed people. This may sometimes be done on purpose, by coincidence, or purely due to a lack of related knowledge. As a result, one might say that right-handed people (explicitly or implicitly) are constantly inducing disabilities in left-handed people, the extent and severity of which is regularly increasing alongside the continuously growing number of manmade objects on the planet. In a world full of tortoises, a rabbit would likely be considered a glorious exception, an all-star athlete perhaps, or something similar. On the other hand however, in a world full of rabbits, a tortoise would most likely be considered a "speed-disabled" individual.

In addition, as an example of environmentally-induced disability consider a dark basement with no windows, cluttered with rubbish. Two people walk in, one of them fully-sighted and one of them blind. If the lights were turned on, then the fully-sighted person would probably be able to move around much faster and with greater ease than the blind person. However, if the lights were to go out, things would eventually reverse, leaving the sighted person with a (temporary) disability. An interesting situation related to this is described in a short story by Herbert George Wells (Wells, 2004) entitled "The Country of the Blind". In this story, a mountaineer accidentally discovers a valley cut off from the rest of the world where an isolated community has remained in which, due to a rare disease, all babies are born blind.

2. Video Games and People with Disabilities

It is likely that there are people who wonder whether people with disabilities actually play video games. The answer is yes. What is more, for many of them games are a key resource for entertainment and socialization as, for example, various mobility problems may render physical activities or just getting together with friends quite difficult or even impossible.

Another typical question relates to how many people with disabilities play video games. A qualitative answer is 'definitely a lot', but no one actually knows since there are currently no related statistics. All the associated available data mostly constitute educated guesses based on national census data related to different types of disabilities and the effect that such disabilities may have on game playing, as well as surveys related to video game players in general (e.g. Yuan et al., 2010; Robinson and Walker, 2010). Such studies clearly highlight the fact that there is a missed multibillion dollar market opportunity related to game accessibility. Yet beyond this, the real question should be "*how many people (with or without disabilities) would actually play games if they were made more accessible?*"

The key disabilities affecting game accessibility are mainly related to:

- *Visual impairments* such as blindness, low vision and color blindness.
- *Motor impairments* such as the absence of limbs or digits, paralysis, a lack of fine control, and instability or pain in the use of fingers, hands, wrists or arms.

- *Hearing impairments* which may range from total deafness to slight loss of hearing.
- *Cognitive impairments* which roughly include difficulties in the performance of mental tasks, ranging from limited and focused problems affecting a very specific cognitive function to severe cases where the individual is unable to carry out the activities of daily life.
- *Speech impairments* such as problems with articulation, the inability to speak loudly or clearly or even the inability to speak at all.
- *Illiteracy* which, whilst not a physical disability *per se*, can have a considerable impact on game accessibility.

Age-related disabilities are frequently referred to as a separate category, but the problems related to age all fall within one or more of the above categories.

At a high level, problems related to game accessibility include (Yuan et al., 2010):

- Providing input.
- Receiving feedback and properly processing and understanding this.
- Determining what to do.

These problems may range in severity from being annoying to making gameplay impossible. For example, providing input through a highly complex device (as most console joysticks are) may be simple for a teenager with no disability but challenging for an adult, difficult for a novice player and impossible for a person with hand-motor impairments. Similarly, receiving visual feedback on a mobile telephone screen may be easy for someone with perfect vision, challenging for an elderly person, difficult for a person with low vision and impossible for someone who is blind.

3. Making Games Accessible

Depending on their accessibility considerations, games can be broadly classified into two categories:

a) "Mainstream" commercial games, targeted at PCs, consoles, mobile phones, on-line platforms etc. that have no particular accessibility considerations.

b) "Special" games developed to be accessible by specific categories of user, such as one-switch games for people with motor-impairments, audio-only games for blind people etc. Such games may be commercial (but usually come from small independent developers) or free, public domain software.

In order for games in the first category to be made accessible, various types of "adaptations" are employed. These may include special (commercial or custom-made) hardware devices, assistive software tools, and – more commonly than you may think – hacking and computer trickery. Additionally, help from another person may be required e.g. to turn on the device, to start/pause the game or to share game control. In essence however, trying to make a game accessible with no related design provisions is somewhat like kissing a frog and hoping that it will become a prince. Typically, only a limited form of accessibility can be achieved, coupled with poor quality of interaction and usability. On top of that, if a game is not purposefully designed to be played using "accessibility features", there is a good chance that it might be made accessible but still not be fun to play.

Creating "special" games is sometimes the only solution, but this approach still has two key drawbacks, since there is:

a) a significant cost associated with the development of high quality accessible games, while the expected return on the investment is rather low as the target user group represents a limited market population; and
b) an apparent risk of social exclusion due to the potential segregation between able and disabled gamers, or even amongst the diverse corpus of disabled gamers.

There are currently no official guidelines or standards nor any world-wide initiatives in the domain of game accessibility. There are also no related governmental or legislative actions. The only (rather outdated) available game accessibility guidelines come from two groups (Ossmann and Miesenberger, 2006):

1. The Game Accessibility Special Interest Group of the International Game Developers Association (IGDA) that in 2004 published a White Paper (IGDA GA-SIG, 2004) concerning game accessibility, including a list of possible approaches for providing accessibility in games. The SIG has also created a "Top Ten" list of things game developers can do to start increasing the accessibility of their games with minimal

effort on their part and without greatly affecting (and perhaps even improving) general game play (IGDA GA-SIG 2006).
2. A group led by the University of Linz and the Norwegian company MediaLT which, on the basis of a set of "Guidelines for the development of entertaining software for people with multiple learning disabilities" published by MediaLT in 2004 (MediaLT, 2004) and the guidelines published by GA-SIG, attempted to create guidelines that adopted the philosophy of the W3C/WAI Web Content Accessibility Guidelines (MediaLT, 2006).

4. The Mainstream Games Industry vs. Game Accessibility

For many years, the mainstream games industry (in a similar manner to the software industry) has been designing games based on the myth of the "average" player. According to this approach, game designers target a fictitious character with a specific range of characteristics which are based on statistical data (or sometimes just assumptions) about the potential players of a game. This results in "one size fits all" games, in which even small design details may lead to the exclusion of large groups of players. Another surprisingly common "design" practice can be summarized by the phrase "build it first and then find what is wrong with it". A game is developed and then a number of evaluation sessions with experts and representative players are run in order to identify problems that need to be fixed. This means of course that only minor details can actually be remedied, since repairing the really big and important design problems at such a late stage is prohibitively expensive.

As a metaphor to illustrate the relationship between the games industry and the aforementioned approaches to game accessibility, if the game industry was a restaurant, it would serve just one type of food; for the sake of this example we will say cheeseburger with fries (Figure 1). In this imaginary restaurant, this is the only meal that anyone can have to eat. No one can have anything else other than this set meal. There can be no variation in even the *type* of burger and the fries are mandatory, not optional. On top of that, the games industry restaurant would require that everyone eats using a specific set of "compatible" plastic utensils. On occasion, some mustard and ketchup *may* be provided to spice up the meal a little.

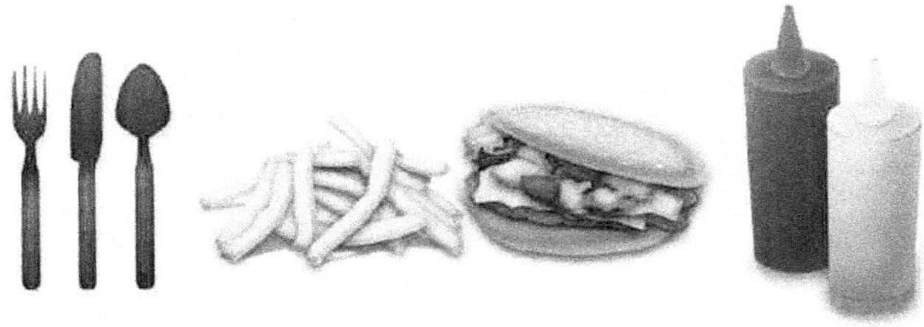

Figure 1: The approach of the mainstream games industry.

To solve this metaphorical problem by following the first approach used for real to create accessible games as mentioned in the previous section (i.e. the adaptation of non-accessible games), some restaurant proprietors may decide to say to their diners that: "Unfortunately we cannot do anything about the food, but we can provide you with a larger variety of utensils that better fit your needs." (Figure 2).

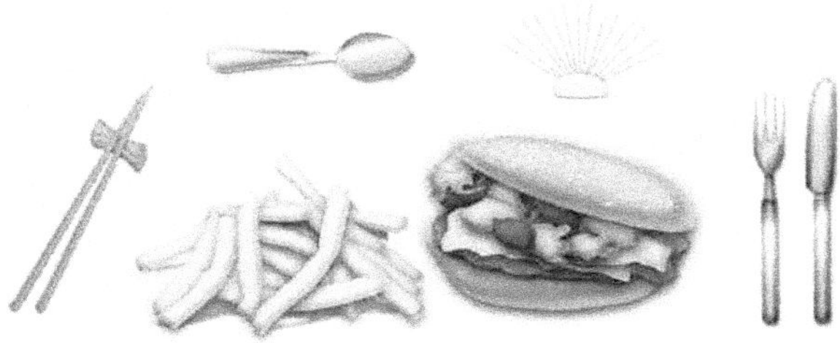

Figure 2: The "adaptation of non-accessible games" approach.

If the second approach to videogame adaptation was followed in the restaurant scenario (i.e. the development of "special" games), then the proprietors would say to their diners that: "We cannot cook something as complex as a cheeseburger, but we can use some of its ingredients (or even others) to create simpler dishes that are appropriate for your diet, so that at least you will not starve. Ah yes, and you can also use any utensils you like!".

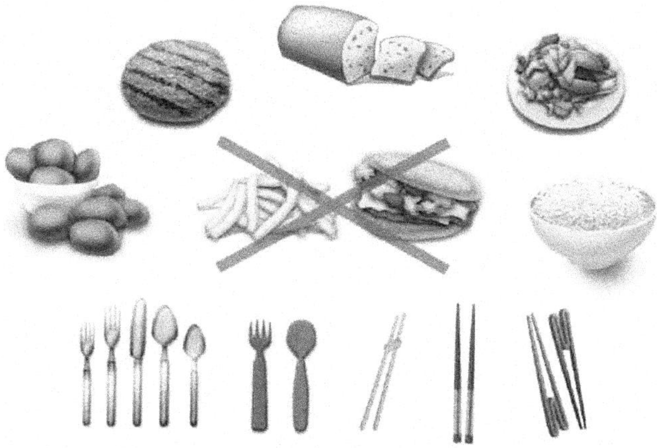

Figure 3: The "development of "special" games" approach.

However, one might ask that since we already have all these various ingredients and utensils to hand, why can we not follow an alternative approach to cooking? Why, for example, can we not mix ingredients in various ways, so that we can serve a much larger assortment of meals, meeting the needs of as many different people as possible? This is the basic philosophy behind the concept of Universally Accessible Games (Figure 4).

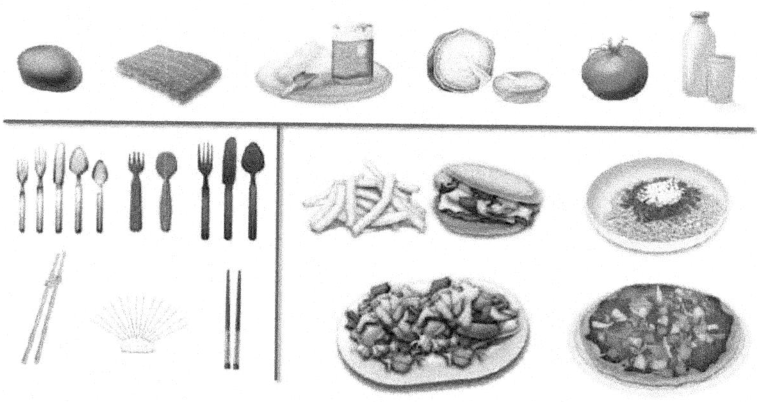

Figure 4: The "Universally Accessible Games" approach.

5. Universally Accessible Games

The Universally Accessible Games approach (Grammenos et al., 2005) constitutes a holistic approach to game accessibility, rooted in the principles of *Design for All* (Stephanidis et al., 1998). Universally Accessible Games (or 'UA-Games') can adapt to the different characteristics of individual gamers and be played concurrently by people with diverse requirements, if possible even while sharing the same gaming device. The global vision is that through UA-Games, *everyone* will be able to play and have fun together, collaborating or competing on an equal basis, irrespective of individual characteristics, preferences and the technologies used. The distinctive characteristic of a Universally Accessible Game is that it can adapt its interface, gameplay and content to best serve the requirements of each gamer, under specific gaming conditions.

5.1 Designing Universally Accessible Games

People most often play games to have "fun", which according to the Merriam-Webster dictionary is "what provides amusement or enjoyment". An important characteristic of fun is that it is highly subjective. Some people find solitaire fun, whilst for others fun involves blasting hordes of heavily-armed aliens. In games, fun typically stems from challenge e.g. beating the clock, gaining money, capturing the flag, shooting the aliens, eating the dots etc. Each and every game has one or more challenges which may be mental, physical or both, and may range from the trivial to the nearly impossible. As with fun, what constitutes a challenge is also highly subjective, so when creating Universally Accessible Games, alternative levels and challenge types are supported (as illustrated in Figure 5).

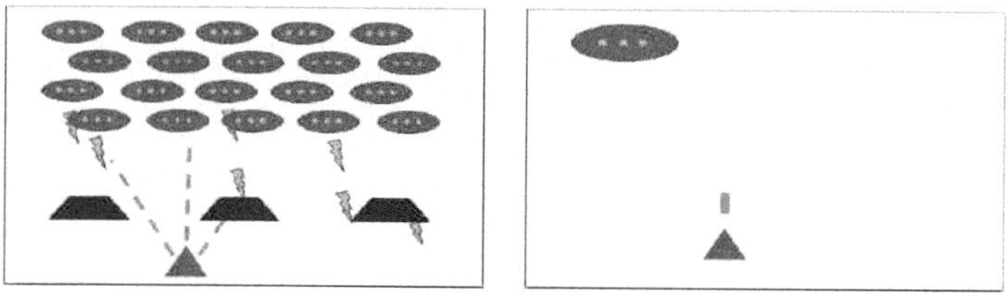

Figure 5: Different people may have a different view on what does and does not constitute a challenging game.

The underlying design philosophy of Universally Accessible Games is that instead of creating a monolithic game for everyone, a "palette" is employed comprising all game elements. Then, depending on the current player, the most suitable elements are selected, with their attributes also appropriately adapted, in order to render a fully personalized version of the game (Figure 6).

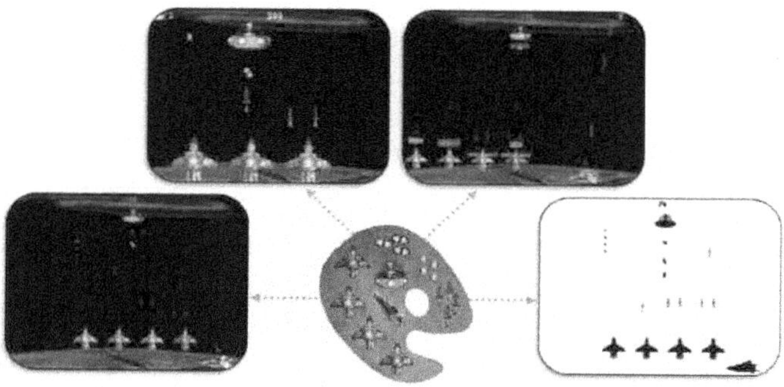

Figure 6: The underlying design philosophy of Universally Accessible Games.

There are 3 complementary tools that a game designer has to make a game accessible to a specific player under specific playing conditions (see Figure 7):

1. *Assistive technologies*: i.e. hardware and software suitable for a specific "disability" which to some extent compensates for it.
2. *Interaction techniques* appropriate for the player's interaction capabilities and preferences. Game designers can work with, and take advantage of, any available assistive technologies.
3. *Content and gameplay adaptation* in order that the game can be rendered in a format that can be optimally perceived and used through the employed assistive technologies and interaction techniques.

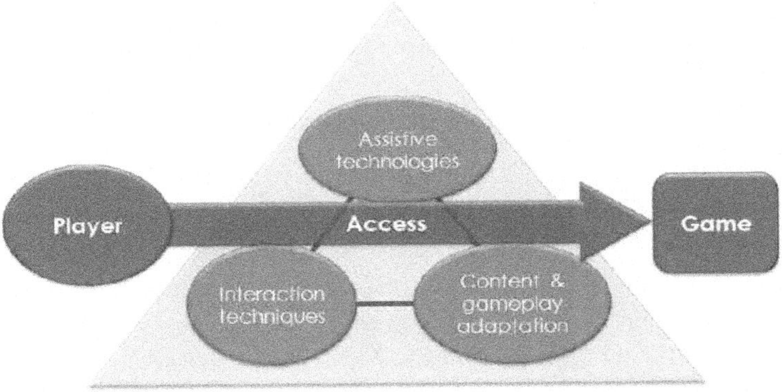

Figure 7: Game accessibility solutions.

If used properly, the right mix of these 3 ingredients can potentially solve any accessibility problem. In fact, although there are several different user categories and contexts of use, they both share many similarities and requirements. For example, a deaf person, someone in a noisy place, or someone playing with muted sound all have the same needs. Thus, when designing for game accessibility, most of the time a single solution is likely to accommodate multiple problems and situations. In other words, designing for game accessibility is much like solving a crossword; the more words you fill in correctly, the more additional ones are "automatically" revealed.

In order to formally design Universally Accessible Games, a structured design method has been suggested, namely the *Unified Design of Universally Accessible Games* (Grammenos et al., 2007). This is based on the Unified User Interface Design method (Savidis and Stephanidis, 2004) and reflects a process-oriented discipline emphasizing abstract task definition with incremental polymorphic physical specialization. The basic steps in applying Unified Design are summarized in Figure 8 (for further details, see Grammenos et al., 2007):

Step 1: Abstract task-based game design

The goal of this first step is to breakdown the high-level tasks performed by people when playing a particular game (irrespectively of the medium they use to play it) as well as the things they do, the things they act on and the things they need to know. In this context, it is essential to focus on the basic logical game activities and constituents, identifying their semantic attributes and relevant regulations independently of the way these can be physically instantiated to be accessible or usable to particular user groups.

Step 2: Polymorphic specialization with design alternatives

Tasks resulting from Step 1 are mapped onto (multiple) low-level physical alternative interactive designs. Potential accessibility barriers for each task when performed by a particular user group are identified and suitable alternative interaction methods and modalities are selected.

Step 3: Analysis of the appropriateness of design alternatives

A matrix is constructed correlating the perceived appropriateness of each selected design alternative to every user attribute by reviewing related literature, using previous knowledge within the field and asking domain experts and representatives of the target user groups.

Step 4: Compatibility analysis among design alternatives

Cases where two or more alternatives are mutually incompatible are identified in order that they can be avoided.

Step 5: Prototyping, usability and accessibility evaluation

Prototypes are constructed, showcasing alternative interactive properties for the different target user groups, and also evaluated with representative end-users in order to assess decisions made at a specific step, or to weigh up the alternatives before committing to them. The outcomes of this step can considerably aid in validating, correcting and updating design decisions, as well as in developing new ideas for improving the accessibility and playability of the final game.

Unified Design is a highly participatory, user-centered and iterative process, as: (a) the direct involvement of several representative end-users (gamers) with diverse characteristics, as well as domain experts (usability, accessibility, gaming, etc.), is promoted throughout the overall lifecycle in order to allow for the continuous assessment of the design outcomes at each step; and (b) it is possible to return to a previous design step in the case that, for instance, more information is required, some design artifacts have to be revisited or the design parameters need to be further specialized.

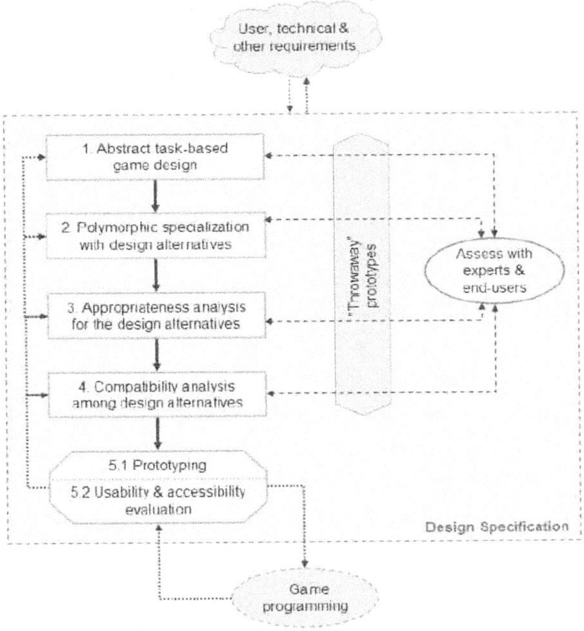

Figure 8: Applying Unified Design to the development of Universally Accessible Games.

Using the Unified Design method, four computer games were created with dual roles: to act both as proofs of concept and as case studies:

1. *UA-Chess:* a universally accessible Web-based chess game (Grammenos et al., 2005).
2. *Access Invaders:* a universally accessible multiplayer and multiplatform version of Space Invaders (Grammenos et al., 2006).
3. *Game Over!:* a universally inaccessible game intended to be used as an educational tool for disseminating and teaching game accessibility guidelines (Grammenos, 2008).
4. *Terrestrial Invaders*: a UA-Game packed with numerous accessibility features, developed in order to create Game Over! (Grammenos, 2008).

A detailed account of the games and their design can be found *Designing Universally Accessible Games* (Grammenos et al., 2009). Two of the games (UA-Chess and Game Over!) have received awards of distinction in recognition of their value in providing universal access.

6. A Far More Challenging Task: Multiplayer Accessible Games

A really challenging task when developing Universally Accessible Games relates to supporting multiplayer sessions where people with diverse (dis)abilities can play a game cooperatively or competitively, being fully aware of each other while at the same time experiencing the game in an optimally adapted way. The idea of Parallel Game Universes (PGUs) (Grammenos, 2006) was conceived as a potential solution. In brief, the suggested approach is to allow each player to play in a different "game universe" and then somehow project each universe onto the other(s). A "game universe" is an instance of the game after it has been adapted to best suit the requirements and needs of a particular gamer playing under particular conditions. In this context, a "transition function" is required in order to translate the events of one universe into the other in a format that is suitable and meaningful in the receptor universe. An important thing to note at this point is that the overall objective is not to recreate everything that exists or occurs in a particular universe in every other one, but just to communicate enough cues in order that the players can cooperate in a successful and enjoyable manner.

Parallel Game Universes are characterized by two key properties: *individualization* (Figure 9) and *balance* (Figure 10).

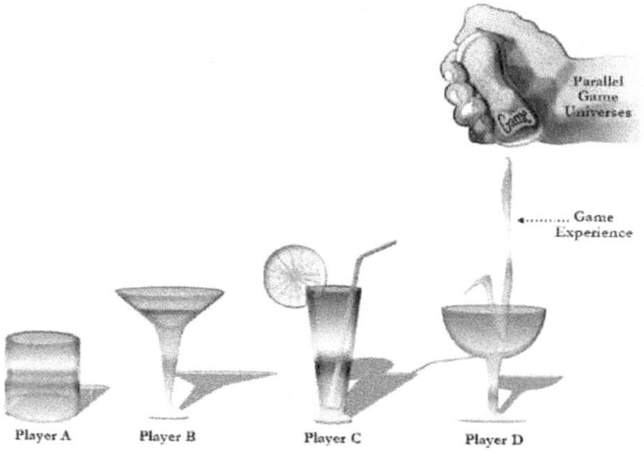

Figure 9: PGUs support *individualization* by extracting the pure essence of games (which is *game experience*) and offering it to the individual players according to their needs and preferences.

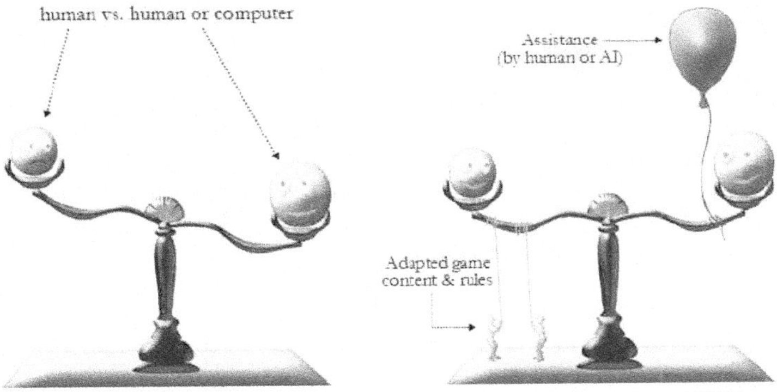

Figure 10: PGUs strive for *balance* by compensating for individual player weaknesses and challenging player strengths, ensuring that opposing forces (player vs. player or player vs. computer) are matched.

As an illustrative example, consider a tennis game between a blind and motor-impaired person (Player A) and a sighted person (Player B) (see Figure 11). In the game universe of Player A, the game is rendered as a simple 2D game resembling the game of Pong augmented with spatial sound. The player's goal is to pinpoint the ball's position in the 2D space using hearing, and to place the bat underneath it using 2 switches. Through the medium of sound, the player can glean additional information such as the opponent's position and the current score. In Players B's universe, the game is represented as a realistic 3D tennis simulation seen from a first-person viewpoint. The player controls an athlete using a gamepad and 8 buttons. In order to hit the ball, Player B must position the athlete correctly in space and also adjust the height and movement of the athlete's arm. The two universes are "synchronized" through a transition function (f_T) which is responsible for translating the 3D positions and speed vectors of the ball and players to 2D, and vice versa.

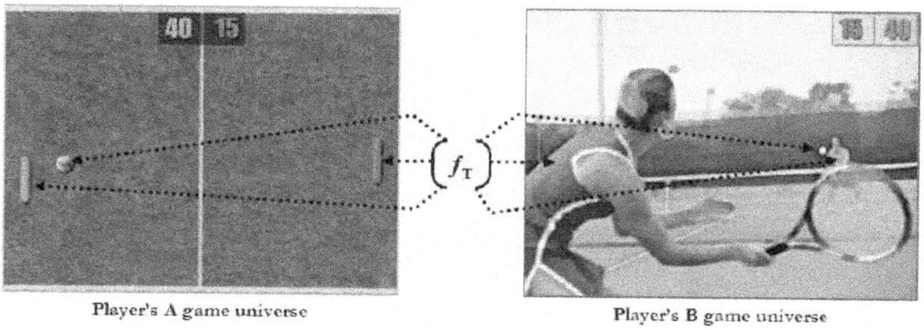

Figure 11: An example of 2 players competing in Parallel Game Universes.

The only limit to what can be achieved through PGUs is the design of the transition functions used. A rather challenging example is presented in (Figure 12), where two distinct transition functions (f_T and g_T) for making a game of chess accessible to people with different mental skills are presented. In this case, the transition functions use AI to transform the complex problem of selecting the best possible move on the chessboard to a much simpler one. Naturally, since the simplified versions of the game may take less moves to finish, a single match in one universe (e.g. on a full chessboard) may correspond to several matches in another (e.g. tic-tac-toe). Clearly in this example the mapping between the two universes is not straightforward due to the significant differences between them. However, the goal of the game designer is not to find a computationally equivalent model for the two universes, but to devise a creative solution so that two players with highly diverse skills and abilities can compete on an equal basis.

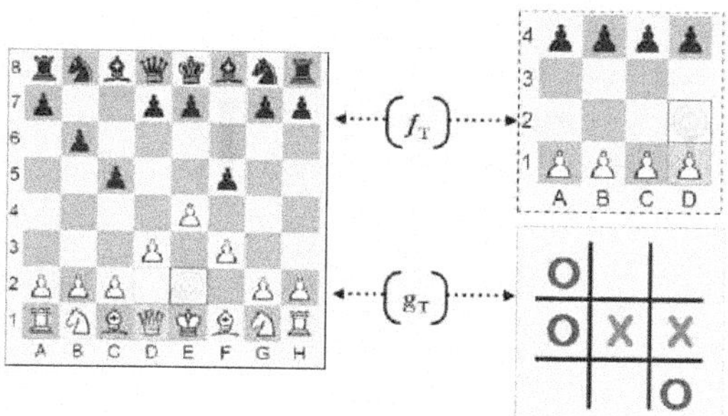

Figure 12: An example of two potential transition functions for making a game of chess accessible to people with different mental skills.

When designing a parallel game universe, the game is "translated" into each player's "language". This task shares a lot of similarities with translating literature (or, better still, poetry) as it is not achieved by following a word-by-word approach. The goal is instead to reproduce the ambience, feelings, images etc. as described by the author and to render the essence of the work, not the actual words originally used. As a result, it is quite likely that different players will not be playing the "very same" game. In such cases, a key question is whether or not this matters. The answer is rather subjective, but what should ultimately matter is that these people are given the chance to enjoy the maximum level of fun and challenge possible when playing a game, without having to compromise or sacrifice their personal gaming experience due to their individual differences.

A typical criticism (usually leveled by 'hardcore' gamers) regarding PGUs is that the proposed mechanisms can potentially be used for cheating, especially in the context of competitive on-line games. Such problems can be easily remedied through appropriate strategies, such as making all activated "game-aids" visible to all players (e.g. auto-shoot, extra health and AI-support) or by giving players the option not to compete against others who have specific aids activated, such as by allowing that they can be invisible (or invulnerable) to each other. The concept can work for any type of game, even for Massively Multiplayer Online Role-Playing Games (MMORPGs) since, in essence, it only affects the way that the game world is perceived by the player and the mechanisms through which the player can (inter)act within it.

6.1 Ambient Intelligence: The Next Big Challenge for Game Accessibility

The vision of Ambient Intelligence sees a future where our surrounding environment is populated by several interoperating computing-embedded devices of different size and with different capabilities, which are interweaved into "the fabric of everyday life" and are indistinguishable from it (Weiser, 1991). In such "intelligent" environments, the way in which people perform everyday tasks is expected to radically change. Multimodal, direct "natural" methods of interaction such as speech, touch and gestures are expected to be widely used, in combination with knowledge about contextual factors such as the profile, preferences and location of the user.

A key problem is that just as Ambient Intelligence Environments will effectively combine the real and the digital worlds, they will also do the same with their (game) accessibility problems, resulting in far more complex design challenges. Two recent related gaming examples are the Nintendo Wii and MS Kinect, which although introducing novel interaction possibilities for many players, simultaneously rendered inaccessible several games that would otherwise be totally accessible if using standard controllers (e.g. by wheelchair users and others who cannot – or do not want to – engage in high-level physical activity).

Such problems are not of course insurmountable. They simply require novel ways of thinking. For example, a possible way of easily achieving accessibility in most of Kinect games is through a "Frankenstein" approach, where a virtual skeleton of a player is assembled using the (real) tracked body parts of multiple real players or even AI e.g. one player is the hands, another the legs and the game AI the head. This approach can be easily implemented and does not require significant computational resources. It may also result in a new type of fun (and social) gaming for everybody.

7. In Summary

Admittedly, the vast majority of accessibility problems found within games are due to the designers' lack of relevant knowledge. In fact, most of the time a game can become highly accessible simply by: (a) taking appropriate decisions at the design stage; and (b) avoiding commonplace design pitfalls. In short, some things that one can do to increase game accessibility include supporting:

- Multiple input devices and techniques.
- Control customization, e.g. adjustable sensitivity, being able to play using fewer or simpler controls (ideally just a single 'button'), avoiding the need for (or providing alternatives to) simultaneous button pressing.
- Adjustable game speed and difficulty, as well as automated user actions e.g. shooting, moving, passing etc.
- Scalability of visuals, including both the text and game elements.
- Alternative color schemes and contrast modes.

- Adjustable visual detail e.g. being able to use simpler (or no) textures, background graphics etc.
- Closed captions and sound visualization.
- Distinct audio control for sound effects, music and speech.
- Game sonification i.e. attaching audio feedback to game events, providing audio descriptions, supporting localised (2D/3D) audio and reading aloud (text and menus). Depending on the user profile and game type, the implementation of game sonification may range from the trivial (e.g. reading aloud text dialogues and playing sound effects), to the extremely difficult (e.g. rendering solely through sound a 3D massively-multiplayer action game).
- Accessible documentation. This might sound obvious, but this is probably one of the most commonly neglected game accessibility guidelines.

All the aforementioned features of game accessibility can be highly beneficial to all players, not just those with disabilities. Some examples are presented in Table 1.

Game accessibility feature	*Beneficiary user group*
Closed captions	– non-native language speakers – anyone playing in a very loud (or quiet, e.g. a library) environment
Customizable controls	– left-handed or single-handed people – people with larger or smaller hands (e.g. children)
Support of alternative I/O devices	– players using alternative input devices such as touchpads, non-standard controllers, etc.
Customizable controls + adjustable speed / difficulty	– novice / casual / tired / young / old players
Scalability of visuals	– players using a small or distant screen
Alternative color schemes / contrast modes	– people playing in a bright environment
Sonification + simple controls	– people playing on the move

Table 1: Examples of how game accessibility features can benefit non-disabled players.

By now it should be quite clear that believing that making games more accessible will only benefit gamers with disabilities is like thinking that having seats on buses only benefits pregnant women or that salad bars in restaurants are just for vegetarians! That said, one should always remember that 'accessibility' is not synonymous with 'usability'. A game may be accessible but very difficult (or boring) to play, as for example when using a virtual keyboard to play a game using 18 keys, most of which must be pressed simultaneously. When various game accessibility features are implemented, there should be relevant testing to ensure that the game still makes sense and is playable and enjoyable after alternative combinations of the available options have been selected.

Beyond the largely low-level and straightforward game accessibility features mentioned previously, there are some more difficult but more effective things that one can do to ensure a higher level of game accessibility. These include:

1. Understanding what game accessibility involves and integrating it in the whole game design lifecycle.
2. Initially designing the game at an abstract level.
3. Creating user interfaces that can support alternative methods of interaction and modalities that can co-exist and co-operate, and are able to adapt to alternative user profiles.
4. Consulting real players from diverse user groups.
5. Following an open and extensible interaction design, so that it will be possible to expand the design to cater for more user categories and use contexts at a later stage.

Regarding the reasons why developers should consider game accessibility, there are two complementary views:

(a) *The Egocentric View*
  - Developers can have games that match their skills and preferences.
  - Developers are not getting any younger, and with age come a number of game accessibility problems.
  - Disability is not an uncommon or exotic disease; permanent or temporary, it can happen to anyone at any time (including the developers and their loved ones), and they will probably still want to play videogames.

(a) *The Exocentric View*
  - The developer's games will be better for ALL players.
  - Developers can broaden their target market, thus making more money.

– Developers can make a lot of people happier.
– Simply, because they can!

For many years, game companies have been fishing in the same pond of gamers, competing mostly through improving their "bait". However, the techniques with which they "hook" their gamers are becoming increasingly homogenous. As the number of "fishes" is limited, in order to catch more, they must vigorously compete with each other. A viable alternative is to turn to the "open sea" of previously undiscovered gamers. In order to do so however, there is a need for more than just fancy "baits" – new "fishing tools" are required, along with appropriately adapted methods and techniques. Currently, the game companies' biggest handicap *(no pun intended)* is that they are missing the *∀-factor* (where '∀' is the mathematical symbol of "the universal quantifier", usually read as 'for All').

In conclusion, (universal) game accessibility:

- *is not* a bug to be fixed;
- *is not* an afterthought or an add-on;
- *is* a design philosophy that can introduce great innovations and eventually lead to groundbreaking games;
- *requires* and forces thinking "outside-the-box" which focuses on the players rather than the technology;
- *supports* game democratization, allowing everyone to play with (or against) one another;
- *is not* about people with disabilities – it *is* about *PEOPLE.*

After all, it does not take extraordinary technology to make extraordinary games – just extraordinary thinking!

Acknowledgments

This paper is based on the author's keynote speech, entitled "Universally Accessible Games and Parallel Game Universes", at the 1st International Conference on Translation and Accessibility in Video Games and Virtual Worlds on 2nd and 3rd December, 2010 at the Universitat Autònoma de Barcelona, Spain. An annotated version of the speech's slides can be downloaded from: <http://www.ua-games.gr/publications.html>.

References

Grammenos, Dimitris. 2006. 'The Theory of Parallel Game Universes: A Paradigm Shift in Multiplayer Gaming and Game Accessibility' in *Gamasutra* (17 August 2006). Online at: Available at: <http://www.gamasutra.com/features/20060817/grammenos_01.shtml> (consulted 20.06.2011).

Grammenos, Dimitris. 2008. 'Game over: Learning by dying' in *Proceedings of the ACM Conference on Human Factors in Computing Systems (CHI 2008)*. New York: ACM.

Grammenos, Dimitris, et al. 2006. 'Access Invaders: Developing a universally accessible action game' in *Computers Helping People with Special Needs. Proceedings of the 10th International Conference (ICCHP 2006)*. Berlin: Springer: 388–395.

Grammenos, Dimitris, Anthony Savidis, and Constantine Stephanidis. 2009. 'Designing universally accessible games' in *Comput. Entertain.* 7(1): 1–29.

Grammenos, Dimitris, Anthony Savidis, and Constantine Stephanidis. 2007. 'Unified Design of Universally Accessible Games' in Stephanidis, Constantine C. (ed.) *Universal Access in Human-Computer Interaction. Applications and Services, Proceedings (Part III) of the 4th International Conference on Universal Access in Human-Computer Interaction (607–616)*. Berlin Heidelberg, Germany: Springer.

Grammenos, Dimitris, Anthony Savidis, and Constantine Stephanidis. 2005. 'UA-Chess: A Universally Accessible Board Game' in *Universal Access in HCI: Exploring New Interaction Environments. Proceedings of the 11th International Conference on Human-Computer Interaction (HCI International 2005)*, vol. 7 (CD-ROM). Mahwah, New Jersey: Lawrence Erlbaum Associates.

IGDA GA-SIG. 2006. 'Game Accessibility Top Ten'. Online at: <http://igda-gasig.org/about-game-accessibility/game-accessibility-top-ten/> (consulted 20.06.2011).

IGDA GA-SIG. 2004. 'White Paper: Accessibility in Games: Motivations and Approaches'. Online at: <http://www.igda.org/accessibility/IGDA_Accessibility_WhitePaper.pdf> (consulted 20.06.2011).

MEDIALT. 2006. 'Guidelines for developing accessible games'. Online at: <http://gameaccess.medialt.no/guide.php> (consulted 20.06.2011).

MEDIALT. 2004. 'Guidelines for the development of entertaining software for people with multiple learning disabilities'. Online at: <http://www.medialt.no/rapport/entertainment_guidelines (consulted 20.06.2011).

Ossmann, Roland and Miesenberger, Klaus 2006. 'Guidelines for the development of accessible computer games' in *Proceedings of the ICCHP 2006*, vol. 4061 of LNCS. Berlin: Springer. 403–406.

Robinson, E. and Walker, S.M. 2010. 'Gaming on a Collision Course: Averting significant revenue loss by making games accessible to older Americans'. Online at: <http://ablegamers.org/publications/Gaming_on_a_Collision_Course-AGF-7128.pdf> (consulted 20.06.2013).

Savidis, Anthony and Stephanidis, Constantine. 2004. 'Unified user interface design: Designing universally accessible interactions' in *Int. J. Interacting with Computers* 16(2): 243–270.

Stephanidis, Constantine, et al. (eds). 1998. 'Toward an Information Society for All: An International R&D Agenda' in *International Journal of Human-Computer Interaction* 10(2): 107–134.

Weiser, Mark. 1991. 'The computer for the 21st Century. Scientific American' in 265, (3 September 1991): 94–104.

Wells, Herbert George. 2004. 'The Country of the Blind, and Other Stories. The Project Gutenberg eBook'. Online at: <http://www.gutenberg.org/ebooks/11870> (consulted 20.06.2013).

Yuan, Bei, Elke Folmer and Frederick C. Harris. 2010. 'Game accessibility: a survey' in *Universal Access in the Information Society 10(1)*: 81–100. Berlin/Heidelberg: Springer. Online at: <http://www.springerlink.com/index/10.1007/s10209-010-0189-5> (consulted 20.06.2013).

Alberto Fernández Costales

Translating Fun for All: Promoting Accessibility in Video Games[1]

This paper analyses the role of translation in the process of creating accessible video games. Within the framework of globalisation and the 'information society', video games have become one of the most popular sources of entertainment for millions of people around the world. Moreover, the 'democratization' of games, triggered by the emergence of casual gaming, has created new videogame users, altering the stereotypical profile of 'hardcore' players. In this context, the concept of accessibility needs revisiting, as the new user types may indicate that accessibility relates not only to people with disabilities. Promoting game accessibility is therefore even more important, due to the diversity and complexity of the 'global audience'. As multimedia and multimodal products, video games need to be adapted and translated into target cultures in order to preserve playability as well as to conform to the expectations of users. The ways translation can contribute to (re)creating video games and the connection between game accessibility and audiovisual translation have not yet been sufficiently discussed. This article supports the idea that translation is a key element to fostering e-inclusion and can contribute to bridging the digital divide.

1. Introduction

The progress of new technologies and the technical advances achieved in recent decades have promoted the status of video games as multidimensional and multimodal products (O'Hagan, 2005, 2007). Currently, video

1 This research is supported by the grant from the Spanish Ministry of Science and Innovation FFI2009-08027, Subtitling for the Deaf and Hard of Hearing and Audio Description: objective tests and future plans, and also by the Catalan Government funds 2009SGR700.

games are played by millions of people and in terms of revenue, the games industry can be compared to other creative giants such as the film and music sectors[2]. Although the field of video games has not been sufficiently explored by academia, from a scientific point of view there are many reasons to support its investigation , especially on the basis of the industry's size, the popularity of video games and the fact that games are an example of human-computer interaction (Newman, 2004: 3).

Research into the field of video games has been approached from the point of view of ludology (Frasca, 2001), and narratology (Eskelinen, 2001; Frasca, 1999) among others, and studies have been published focusing on the localisation process and workflows (Chandler, 2005; Chandler and Deming, 2011). As far as Translation Studies is concerned, video games have been investigated by several scholars (Bernal, 2006, 2007, 2008 & 2009; Mangiron, 2004, 2007 and 2011, Mangiron and O'Hagan, 2006; Muñoz Sánchez, 2008; O'Hagan 2007 and 2009). Today, the study of video games can be regarded as an emerging academic field on the basis of the number of workshops and international conferences at which the topic is being addressed. Furthermore, the new movement-sensitive devices (such as the Wiimote, Kinect and PlayStation Move), the growth of online gaming, and the emergence of so-called 'casual'[3] titles (i. e. games which cater not only for established 'hardcore' players but also for potential newcomers) need contextualizing in relation to the framework of a strategy of 'democratization' in the game industry (Mangiron, 2010). This tendency towards democratization or socialization aims to attract more players and reach new user profiles. For instance, some devices such as Nintendo DS and 3DS have launched specific promotional campaigns targeting prospective players over 50 years old and people who do not regularly play video games (or even those who have never played games at all).

Video games have undergone a remarkable evolution not only from a technical point of view but also in relation to narrative and artistic issues. The comparison between the first rudimentary interfaces launched in the

2 According to The Entertainment Software Association, the game industry generated more than $ 25 billion in 2010. Online at: <http://www.theesa.com/facts/> (Consulted 07.01.2012).

3 The term 'casual' is a rather controversial concept, but it is commonly accepted that "the idea of casual games has appeared specifically as a contrast to the idea that video games could only be made for a hardcore game audience" (Juul, 2010: 26).

1960s and the film-like stories that dedicated gamers play today supports the idea that modern video games can be considered to be an art form rather than solely an entertainment product. In this regard, the introduction of additional audiovisual features has allowed companies to create more compelling and complex stories: besides the aesthetic development of the games (resulting from more advanced graphics capabilities), the addition of cut-scenes, voices, dialogues and soundtracks has opened up new horizons, since these developments allow the makers of the games to achieve more realistic gameplay, giving players a better game experience. In a nutshell, video games have become multimodal and multidimensional products (O'Hagan, 2005 and 2007) and the player's interaction and engagement with the game has clearly been fostered. The concept of interaction is a key element to be addressed herein, as the compelling and well-rounded stories of modern titles not only promote a fluent 'dialogue' between the player and the game, but also permit users from around the world to interact together in competitions and open games in the fashionable online modes.

The collective use of video games must be studied within the framework of the Web 2.0, a phenomenon that has largely changed the way people communicate in the information society. The evolution of the Internet towards an interactive model where users can edit, create and upload their own content has had an enormous ripple effect across wider society. In the context of interactive entertainment, online gaming has become more and more popular, and the 'collective wisdom' of the crowd has begun to operate within the games industry, with people modifying and localising games on a voluntarily basis by means of "romhacking" (Muñoz Sánchez, 2008). Similarly, games have become portable and multiplatform, with people no longer solely using their PCs, laptops or consoles to play games, but also their mobile phones. This platform has increased its market share dramatically, with 55% of gamers currently playing video games on their phones or handheld devices[4].

Globalisation and the technical progress of recent decades have allowed most sections of society to access new technologies. In this context, certain concepts such as usability and accessibility have become more rel-

4 Source: The Entertainment Software Association. <http://www.theesa.com/facts/>. Consulted 07.01.2012.

evant in order to guarantee that products can reach the largest possible audience. The key question to be addressed here is this: how accessible are video games? In recent years, institutions such as the World Wide Web Consortium (W3C) have promoted standards for the development of websites that are accessible to everybody. In the case of the games industry, the International Game Developers Association (IGDA) introduced the Game Accessibility Special Interest Group[5] in order to raise game accessibility awareness and contribute to making these products truly available to most people.

Although progress has been made in this field, the technical sophistication of video games and the evolution achieved regarding playability is somewhat overshadowed by the lack of accessibility of many titles, which do not usually take into account people with disabilities or special needs. Most games are not accessible to people with a visual impairment, since they largely rely on graphic and aesthetic elements that these users cannot fully enjoy. On the other hand, the audio component of games also poses barriers to accessibility for the deaf and hard of hearing, as in modern titles sounds and special effects may provide relevant clues or important information about one's progress in the game: for instance, in First Person Shooters, players can establish the enemy position by listening to the sound of gunfire coming from a particular direction (Tavinor, 2009: 77; Mangiron, 2011). Therefore, the introduction of subtitles for environmental sounds can help improve the accessibility of many titles.

The issue of accessibility in video games can be approached from the point of view of Translation Studies by analysing how translation can contribute to making games more accessible. However, is translation actually capable of achieving this objective? As we explain in this paper, the answer to this question is not straightforward and largely depends on what we understand as 'accessibility'. Translation can contribute to granting access to video games to more players by allowing them to play in their own language, thus overcoming linguistic and cultural barriers. In this sense, this paper has been written from a socio-cultural approach to translation, and takes into account issues of reception rather than only of language transfer. As regards interactive multimedia products, translation has to be considered as a multidimensional and multimodal concept: a multidisciplinary approach is required in order to face the challenges of the

5 IGDA Game Accessibility website: <http://igda-gasig.org>. Consulted 07.01.2012.

technological revolution. In other words, the interaction amongst several disciplines (such as Computer Engineering, Game Design and Translation Studies) is a fruitful strategy for shedding light on the study of multimedia products.

Besides localisation, audiovisual translation can provide theoretical grounds for the study of accessibility in the field of electronic entertainment (Orero, 2005b; Tercedor, 2005). In fact, there are insufficient studies on voiceover, dubbing, closed captioning, subtitling and lip-sync techniques in video games. Similarly, the concept of accessibility goes beyond simply making a product accessible for the disabled, as arguably language or cultural barriers can also restrict the target audience of a video game. Therefore, accessibility must be addressed from the point of view of Translation Studies (as well as other disciplines such as Game Design) in order to estimate how the adaptation of textual, non-verbal, semiotic and cultural elements can contribute to enabling the access to video games of a broader audience (not only people with special needs, but also players over 65 and those who speak another language). In this sense, this paper relies on the hypothesis that accessibility is an 'open' concept that in the era of technology and information is constantly evolving to include new user types. For instance, accessibility can also refer to the design of video games for younger and older people (Mangiron, 2011: 56). Therefore, besides considering the role of translation in making games accessible to people with special needs, the article also addresses the adaptation of games for the wider audience and the importance of translation for e-inclusion and bridging the digital divide.

2. Fun for All: Casual Gaming and the Democratization of the Games Industry

Besides the technical evolution of video games and the introduction of online modes of play, the games industry has seen the emergence of a fashionable and appealing phenomenon: casual gaming. Traditionally, playing video games was an activity restricted to a particular group of users, and specific genres such as First Person Shooters, Role Playing Games and Survival Horror were followed and enjoyed by a community of so-called 'hardcore gamers'. However, some companies (particularly Nintendo) investigated the possibility of enlarging the scope of video games and designing their products to cater for a wider audience, instead of focusing solely on 'traditional gamers'. By creating more familiar and interactive games and consoles, electronic entertainment companies provided the basis for a new type of video game. This strategy aimed to reach a wider audience who would play video games not primarily as a hobby, but as an occasional or casual activity. In addition, games catering for casual players were designed to be played collaboratively, and as such multiplayer modes were optimized and encouraged in order to be enjoyed by a group of people rather than by individuals playing alone.

More than simply a fashion trend, casual gaming is here to stay and has already revolutionized the games industry by introducing more players into the arena. The launch of more user-friendly devices such as the Wii or Nintendo DS aimed to attract new gamer profiles, effectively enlarging the number of people playing games: these new devices (supported by huge marketing campaigns) appeal to people who have never played video games by underlining their ease of use, the simplicity of their titles and their social and/or collective nature.

This may have contributed to the changes in the video game consumer profile, with typical gamers no longer being male teenagers or digital natives (i.e. users of new technologies who have been to interacting with computers since childhood). According to statistics provided by The Entertainment Software Association[6], the average age of game players in the US market is 34. Interestingly, 26% of US citizens over 50 play video games, and women account for 40% of total players. Indeed, reports from the International Software Federation of Europe confirm that there are relevant gender differences relating to players' favourite video game hardware. Nintendo's systems are preferred by women and the Xbox 360 is the most 'male oriented' console (ISFE, 2010). This is probably due to the current catalogue of titles available for both systems, which in the case of the Wii are more casual (with many Mario Bros games and family-oriented titles) and rather more 'hardcore' in the case of the Xbox 360 (where First Person Shooters, fighting games and other action titles rule the roost).

6 Updated information and statistics can be found at: <http://www.theesa.com/facts/index.asp>. Consulted 07.01.2012.

As regards the 'social' nature of video games, according to the Interactive Software Federation in Europe, six out of every ten gamers agree that playing with friends is one of the most relevant motivations for playing a video game[7]. In addition, recent surveys show that 58% of European gamers play video games with their children (ISFE, 2010). These figures support the argument that the recent tendency in the game industry has been to focus on reaching a wider audience with the development of more user-friendly devices. In addition, the ubiquity of the Web is also being exploited, as more than 200 million people play Internet-based video games on a regular basis via smart phones, social networks and dedicated websites[8].

It can be concluded that the strategy to foster the consumption of casual games has proven remarkably successful, as it has effectively met the objective of broadening the video games industry's target audience. Although these types of games are usually frowned upon by more 'hardcore' players, who regard this tendency as a minor and temporary fashion, the fact is that casual gaming has contributed to the democratization of the industry (Mangiron, 2010), since video games nowadays cater for all different kinds of members of society. Besides the creation of appealing stories and topics suitable for the whole family, casual games also rely on the design of usable (i. e. easy to use, intuitive and simple) and learnable video games. This does not mean that games are necessarily easier to complete, but that newcomers to the field will be able to interact with the technology in a simple and fluent way: sensitive-control systems and pads are a good example of this, in contrast for example to the complex multi-function joysticks used for playing flight simulators.

Arguably, video games must be accessible to all players in order to guarantee that this process of democratization is effectively implemented. When designing new products and targeting new groups and types of users, game developers need to take into account not only disability, but also the needs of people over 65, as will be commented on in the next section.

7 Statistics are available at: <http://www.isfe-eu.org/>. Consulted 07.01.2013.

8 Source: <http://www.casualgamesassociation.org/news.php>. Consulted 07.01.2013.

3. Accessibility revisited

Accessibility is an issue relevant today, as technical developments have enabled millions of people to access information in real time and from any part of the world. With this situation, people with disabilities and those with special needs must be taken into account, as access to information is a Human Right that needs protecting (United Nations, 2006: 3). Within the framework of our technology-driven society, accessibility should be a major concern in order to allow disabled users to read, edit and interact with digital and multimedia products.

The study of accessibility is gaining recognition and institutions are becoming more involved in granting access to multimedia and digital information (e.g. the World Wide Web Consortium[9], the European Design for All eAccessibility Network[10] and the Center for Universal Design in Education[11]). In this sense, major achievements can be observed in the field of audiovisual translation, with the development of special regulations, the creation of specific institutions (e.g. the Spanish Centre for Subtitling and Audio Description) and the commitment of public and private TV channels to promote subtitling, audio description and closed captioning (Díaz-Cintas, 2005; Orero, 2005b). Unfortunately, it seems that accessibility in video games is less developed than is the case for television, films and DVDs.

Accessibility is a multi-faceted concept that can be approached from different disciplines. As far as Translation Studies is concerned, accessibility has been addressed from the point of view of audiovisual translation (Orero et al., 2007), where significant advances have been made in the fields of audio description and subtitling for the deaf and hard of hearing (Díaz-Cintas, 2001 and 2003; Neves, 2005). Indeed, it can be said that research into audiovisual translation is contributing to the development of more accessible media (Díaz-Cintas, 2008; Franco et al., 2010; Gambier, 2004; Mangiron, 2011; Orero, 2005a and 2011), and the number of con-

9 The Web Accessibility Initiative (WAI) of the World Wide Web Consortium promotes standards and guidelines regarding web accessibility: <http://www.w3.org/WAI>. Consulted 11.01.2012.

10 <http://www.edean.org>. Consulted 11.01.2012.

11 <http://www.washington.edu/doit/CUDE>. Consulted 11.01.2012.

ferences devoted to the field together with the papers on the relevance on training in media accessibility (Badia and Matamala, 2007; Orero, 2005b) suggest that this is not a temporary fashion but a real need that must be addressed by scholars and researchers.

In the context of information sciences, accessibility has particularly been explored in the case of the Web. However, for the purposes of this paper, it is worth commenting on the definition of game accessibility provided by the GA SIG from the International Game Developer Association:

> Game Accessibility can be defined as the ability to play a game even when functioning under limiting conditions. Limiting conditions can be functional limitations, or disabilities – such as blindness, deafness, or mobility limitations (IGDA 2004).

As follows from this definition, accessibility is usually regarded as a process aimed at guaranteeing the access of disabled people to information. However, due to the increasing number of users accessing digital products and their diverse profiles, it can be argued that this definition of accessibility is not sufficiently broad. If we consider language to be potential a barrier in the current multilingual society, accessibility should also take into account those people who are unable to speak foreign languages. Providing information in several languages can therefore be included within the framework of 'Design for All', as it promotes e-inclusion and social inclusion by allowing certain types of user to enjoy technology in their own language. In this regard, this paper supports the idea that accessibility is a broad term in the sense in which it is described by Yves Gambier:

> The issue of accessibility is, however, not merely a question of providing for special visual, auditory, motor or cognitive needs; such a view of the issue is far too restrictive in the light of the digital divide, income-related differences in Internet use, and the exclusion of certain sectors of society from access to information. Accessibility means that AV or electronic products and services must be available to all users, irrespective of issues such as where they live, their level of experience, their physical and mental capacity, or the configuration of their computer. Accessibility is not just an issue for the disabled: it does not only mean a barrier-free situation; it also means that services are available and that information is provided and easy to understand. (Gambier, 2006: 4).

Accessibility can be understood as an 'open' concept which needs to evolve and adapt to new user types. In this sense, several accessibilities can be observed, two of which are of particular relevance to the scope of this paper. Firstly, accessibility is a tool to grant access to people with disabilities, and secondly accessibility is a concept related to e-inclusion which aims to make products available to a wide audience. Although universal accessibility in video games might not be a feasible objective, the use of adaptive technology can make some games suitable for specific groups and translation can increase the number of people (both with and without disabilities) who can fully enjoy a video game.

The possible evolution of accessibility towards a more "inclusive" or broader concept is due to the development and user diversity of current technology. As has been commented on in the previous section, the spectrum of players engaging in video games has dramatically broadened in recent years. Casual gaming and online modes have contributed to a rapid increase in the number of people playing games and a clear diversification has been produced in the target audience. As games become more popular amongst casual gamers, the profile of the typical player has become blurred. One of the most interesting changes is the fact that people over 50 have been attracted to playing certain types of games; in other words, players are becoming older. In this context of change, language issues should also be considered when studying accessibility, as there are a large number of people who do not speak any foreign languages. For instance, according to a survey conducted by the European Commission, in the case of Spain, 56% of the population do not speak any foreign language (European Commission, 2006). Furthermore, Spain is not an isolated case, as the citizens of other EU Member States such as Portugal, Italy, Hungary, Ireland and the UK also show similar percentages (in all the cases greater than 50%).

These figures suggest that localising a video game into the so-called FIGS languages (French, Italian, German and Spanish) is not enough to grant access to information, especially in the framework of the European Economic Area, where multilingualism is on the radar of regional, national and international institutions. In this regard, the use of video games as a tool with which to learn foreign languages is an interesting research avenue that has not yet been sufficiently investigated, although it falls beyond the scope of this paper. However, it is worth mentioning that the needs of people who are unable to learn or speak a foreign language should also be included in the concept of 'accessibility' in order to guarantee the universal access to digital information. In this sense, the translation of video games into more languages (including lesser-spoken ones) is something that should be encouraged, as people should be able to enjoy multimedia products in their mother tongue.

4. Translation as a Tool to Enhance Video Game Accessibility

It can be argued that the translation of audiovisual material into different languages is required in a globalised world. Initiatives such as the programmes 'Media' (which finances the subtitling and dubbing of European films to be distributed and sold in the European Union) and 'Culture' (which promotes the translation of European writers into different official languages) are intended to encourage cultural and linguistic diversity, at the same time granting access to multimedia products to a broader audience. Although there are no similar initiatives in the field of video games, it could be said that translating titles into more languages would be beneficial in a multicultural context and would contribute to enhanced accessibility.

An issue that must be raised at this point is the possible relevance of translation for improving playability. As concluded by Mangiron and O'Hagan (2006), translation (or more specifically, transcreation i.e. the carte blanche or freedom of localisers and translators to adapt a videogame to the target locale, modifying cultural or linguistic elements to make the game suitable for the desired user group) is an element key to preserving the game experience for the target audience. Empirical research needs to be conducted into the physiological response of players and the possible differences between playing a game in one's mother tongue and in a foreign language. In order to solve the challenges of transferring a product from the source into the target culture, translation can effectively contribute to conveying and adapting ideas and even emotions for the target audience.

Since accessibility is aimed at promoting the integration of individuals in society (Orero, 2005b), granting users access to digital content in their own language is obligatory in the context of a multilingual and multicultural community. The role of translation can be reinforced and enhanced if it is regarded as a tool intended to contribute to the construction of a more accessible and equal society.

Making accessible games is a process aimed at overcoming the potential barriers experienced by people with disabilities and enabling them to play a video game. In this regard, four different categories of accessible games can be established: games for the visually impaired, games for people with auditory disabilities, games for people with mobility-impairments, and games for users with cognitive disabilities. This taxonomy can be subsequently broken down into several sub-categories: for example, designing a game for the visually impaired implies adapting titles both for blind people but also for gamers with poor vision or colour blindness. Although the elements which affect game accessibility are manifold, this section will only focus on the two areas where translation is particularly relevant for meeting the objective of making games suitably accessible: subtitling and dubbing. The way in which game accessibility can be improved for people with special needs will be commented upon in sections 4.1 and 4.2. In addition, the role of languages and translation for improving e-inclusion and bridging the digital divide will also be discussed (section 4.3).

4.1 Visually Impaired Gamers

The technical development of video games and the optimization of graphics capabilities led to a shift from computer games to video games, reducing the amount of text and focusing on visual elements. Games therefore tend to be less and less accessible for people who are blind, have poor vision or colour blindness; these users have two basic alternatives to mainstream titles: audio games and text-based games which use adaptive technology.

Audio games are titles specifically designed for users with visual disabilities. In a nutshell, audio or auditory games are "computer games that feature complete auditory interfaces, so they can be played without the use of graphics" (Friberg and Gärdenfors, 2004). As such, the visual component is not an important issue in the development of audio games, since these titles are intended to be enjoyed using a player's listening capability. Thus, the audio component, the narrative techniques and the recording of voices and sound effects are the main tools and resources for making a game playable and accessible for blind users.

As far as audio games are concerned, translation is an essential element when localizing the title into other languages. Naturally, the way to adapt audio games in which subtitles cannot be displayed and all the information has to be conveyed through recorded voices is by dubbing (voiceover is not used, as the overlapping of voices could distract users and disrupt the game experience). In this sense, it is worth mentioning that the translation and localisation of video games is included in the framework of constrained translation (Mayoral et al., 1988), in which the adaptation of contents must be carried out whilst taking into account the particular contextual restrictions or characteristics. In the case of video games, constraints refer to the limits 'imposed' by the screen (in particular, the reduced text size in menus etc.) and also to the most common challenges of subtitling, such as reading speed and the transferring of non-verbal elements and nuances such as accents and intonation (Díaz-Cintas, 1998).

It is also worth mentioning that translation into other languages is likely to require a great deal of creativity, cultural adaptation and perhaps even transcreation, which can all improve the playability for users playing a localized version of a game. In this sense, transcreation can improve the playability for users who are playing a localized version of a game. If the game has not been properly adapted to a particular locale, part of the title's essence may be lost in translation.

Text based computer games such as text adventures (and even graphic adventures) can be played using screen readers. Unfortunately, this alternative is only available to PC users, as for the current generation of consoles there are no software applications that can be used to read the information included in these games. In any case, it is interesting that most players prefer to enjoy the experience of games with the original voices or dialogues, as the synthesized sound of screen readers is rather unrealistic and impacts negatively on the game experience of visually impaired people (as concluded following a survey conducted by France, 2007). The same report also states that the three preferred genres for people with visual disabilities are RPGs, Survival Horrors and Adventure Games. As these types of games rely on heavily-driven narrative techniques, the creation of realistic and compelling environments is the key issue to be addressed when developing this kind of title. In order to achieve this objective and meet the expectations of the visually impaired, we can hypothesize that language and textual features are particularly relevant in these genres (having an auditory input can improve playability for users with visual disabilities and thus many of them will prefer this kind of game).

On the other hand, mainstream titles can also be played by people with visual disabilities using adaptive technologies. In this regard, certain series such as *Grand Theft Auto* (1998–), *Soul Calibur* (2002–), *Tekken* (1995-) and other fighting and action games are played by blind people thanks to the auditory feedback provided by the games[12]. However, the number of mainstream titles able to meet this criterion is rather limited.

12 Source: <http://www.game-accessibility.com/index.php?pagefile=visual>. Consulted 27.01.2012.

The main reason is the heavy dependence of modern games on powerful and impressive graphics that are intended to be as visually realistic as possible. Console games are therefore becoming increasingly inaccessible for the visually impaired. In this regard, the narration of all menus, game options and in-game elements seems to be the most suitable solution for allowing the visually impaired to enjoy video games at an acceptable level. It is a feature that could be incorporated into many (mainstream) games that could subsequently be localised into different languages, improving game accessibility. Unfortunately, to the best of the author's knowledge, there are still no games which include audio description.

4.2 Hearing Impaired Gamers

Users with auditory disabilities can interact with video games through the visual interface. As has been mentioned previously, the development of modern titles is clearly aimed at achieving the most powerful and impressive visual effects. However, in spite of enjoying the aesthetic and artistic components of games, the introduction of technically advanced features has also restricted their accessibility for the deaf and hard of hearing. As far as sound effects are concerned, video games have evolved from the primitive beeps of the first titles to the cinematic soundtracks of modern games, which also include real voices recorded by actors performing in-game dialogues (Chandler, 2005: 186; Mangiron and O'Hagan, 2006). Beyond the technical issues, this represents a landmark in video games development, as the introduction of real voices and music can be used as a narrative technique aimed at creating more complex and deeper story lines. At this point, it is worth mentioning how the automatic sequences of cut-scenes (which show the progress of the story) have been increasingly used in recent titles. In this regard, lip-sync techniques are increasingly being used in the development of mainstream games, and subtitles should be synchronized with the actual game dialogues according to the basic rules of audiovisual translation (Agost and Chaume, 2001; Díaz-Cintas, 2001: 41). However, it is also worth mentioning that a gap exists between audiovisual translation and game localisation, as these rules do not often apply when adapting a title to a particular locale.

The creation of narrative-driven plots and more complex games also may pose a problem for users with auditory disabilities, as for many titles the audio component is a key element in the story. It could be argued that some genres (such as RPGs) are more narrative-driven than others (Mangiron, 2004), and the audio element of video games may be required in order to fully experience the story with which the player is interacting. For example, cut scenes which include audio cannot be fully enjoyed by the deaf and hard of hearing if subtitles are not provided.

In this context, subtitling is the main tool used to allow the deaf and hard of hearing to read the dialogues and voices of a game's characters. Most modern titles include the option to activate intralingual subtitles where dialogues are shown in the player's mother tongue. Furthermore, many games also allow players to select the original language of the game (or even other languages) – a useful tool for learning a foreign language. For instance, *Assassin's Creed II* (2009) and *Assassin's Creed Brotherhood* (2010) can both be played in Italian by a Spanish or a French player willing to learn the language (or simply by users trying to feel more immersed in the Italian setting and atmosphere of both games).

Even though the possibility of displaying subtitles is a major breakthrough in promoting game accessibility and is very much appreciated by players with hearing impairments, there are some issues that need to be improved regarding the subtitling of games. The differences between interlingual and intralingual subtitles in video games are limited to the language that is displayed on the screen (i. e. the source or target language). However, there are no SDH (Subtitles for the Deaf and Hard of Hearing), so subtitles are created for all users and not for players with special needs. This means that specific features (such as ambient noises) are not normally included in the subtitling. It can therefore be said that the needs of people with disabilities and the time and space restrictions on subtitles are not afforded the same level of importance as in the translation of films or DVDs (Díaz-Cintas, 2003: 146–156).

Video game subtitles seem only to address people without disabilities who activate them in certain situations (for example if playing games on a mobile phone in a public place when they may prefer to activate subtitles and turn all sound options off). In fact, it can be said that in many cases, some basic elements are not taken into account in audiovisual translation: for instance, some games do not include any reference on the screen to the character who is actually uttering the sentence (although there are some titles which include the name or initials, a symbol or even the face of the character next to the subtitles). Therefore, players who are unable to listen to spoken voices and dialogues cannot distinguish which character is 'speaking' if it is not clearly shown in the scene. This lack of attribution can be observed in many blockbusters, such as *Allan Wake* (2010), *Assassin's Creed* (2007) and *Gears of War* (2006).

In addition, subtitles usually only transcribe the dialogues of the game, forgetting the rest of the sound effects or the music. Although in some genres such as racing or sports games this can be considered a minor inconvenience, in other cases it may become a serious barrier, since it will hamper usability as well as accessibility and gamers will face additional challenges to advancing in the story. This is the case for Survival Horrors such as *Silent Hill 4* (2004) and *Resident Evil 5* (2010), where music plays a special role in creating the atmosphere and suddenly changes to warn the player about the proximity of enemies or dangerous situations. Similarly, First Person / Tactical Shooters such as *Medal of Honor* (2011) and *Call of Duty* (2011) can become extremely difficult to play if users are cannot listen to the enemy fire (or noises), as they will be unable to notice the presence of rival units until they are physically evident on screen (and it is consequently too late for the player to react).

A remarkable exception to the general standard within the industry was the subtitling system used in the original US (PC) version of the mainstream title *Dragon Age: Origins*[13] (2009). In this game, it is possible to activate 'ambient noise subtitles' (normally referred to as closed captioning) which also relate the sounds of the game in addition to the traditional subtitles which are displayed by default in cut-scenes and in the rest of the game. Although ambient noise is not essential to the gameplay, this option contributes towards eliminating any possible disadvantage for the deaf and hard of hearing. Also, in the combat scenes of this game, the character controlled by the player automatically draws his/her weapons, alerting those users who cannot listen to the music or the voices of approaching enemies. However, it is necessary to underline that this option (i.e. closed captioning) was only available in the original game and not in the localised versions, which reduces their accessibility.

Although titles offering closed captioning are still scarce[14], some modern games include partial or full closed captioning. Examples of such titles

13 This game was awarded the 2009 Accessible Game of the Year by the website <www.ablegamers.com>.

14 Interestingly, some groups have been created in order to create mods which include closed captioning. One of the most relevant initiatives is Games[CC], created by Reid Kimball. <http://gamescc.rbkdesign.com/>. Consulted 27.01.2012.

are the PC version of *Crysis* (2007), *Left 4 Dead 2* (2009) and the *Metal Gear Solid* series (1998–). It is noteworthy that the International Game Accessibility Association ranks the introduction of closed captioning for all dialogues in third position on its list of the top ten game accessibility features (after allowing controller reconfiguration for improved comfort and providing alternative controller support)[15].

4.3 Media Accessibility for All: Reducing the Digital Divide

Video games have gained momentum, and the impressive sales figures of the games industry together with the growing number of users support this argument. The emergence of casual gaming has stressed the relevance of accessibility due to the diversity of player profiles and the complexity of modern titles. In the field of multimedia products, accessibility also concerns elderly people using technology and even users with certain degenerative diseases (such as Alzheimer's). Beyond the possible benefits video games could have within educational settings (e.g. for learning foreign languages), people tend to prefer using technology in their own language. As such, localising games from English or Japanese into other languages is a necessity in order to maintain the game experience for mainstream players. Hence, translation should be regarded as a necessary step in the road towards accessibility; otherwise, the expectations of many of the potential users will not be met and they could be excluded from the information society. Indeed, e-inclusion is intended to bring users and technology together by "promoting literacy and cultural diversity through the use of ICT" (Tercedor, 2010). As has been mentioned previously, if we understand accessibility in the broadest terms, it can also apply to people without a physical or cognitive impairment. In the current multimedia society, new technology user types and profiles have arisen, and these people should also be granted the right to access digital products in their own language. Research into audiovisual translation and game accessibility could be especially relevant in the case of those countries which have a high ratio of elderly people unable to speak a foreign language, as translation can clearly soften the learning curve within video

15 Source: <http://igda-gasig.org/about-game-accessibility/game-accessibility-top-ten/>. Consulted 07.01.2012.

games, thus enabling access to a wider range of players and promoting e-inclusion.

In this sense, allowing people to play in their native language can be related to e-inclusion, as it allows access to digital entertainment for a wider audience. Although it cannot be directly compared to other practices intended to improve accessibility (such as using different font sizes and colour contrasts for the visually impaired or enabling text messaging in online gaming for the deaf and hard of hearing), translation can also promote accessibility if we understand its concept to be guaranteeing access to information to the largest number of people possible. In this context, people over 65 or users who cannot speak a foreign language can also access audiovisual products in a similar fashion to DVDs or films. In the case of video games, translation can to some extent promote accessibility and e-inclusion. To illustrate this with an example, most European versions of the previously mentioned Dragon Age: Origins are not fully accessible to certain users because sound effects are not included in the localized subtitles (closed captioning is not available). Although players can activate subtitles for cutscenes, in most locales the option of 'ambient noise subtitles' is not available. In addition, the voices of the game have not been dubbed (only interlingual subtitles are provided), making these versions less accessible than the original, as visually impaired players do not have access in their mother tongue to elements essential to the game such as the dialogues and voices.

Subtitling, dubbing and closed captioning are the main translation activities involved in the development of accessible games. Applied research into audiovisual translation (and Translation Studies in general) will be able to offer new insights and promote best practices not only from an academic, professional and even technical point of view, but also from a socio-cultural perspective.

5. Conclusions

Accessibility is a broad term that should not be solely concerned with the way in which people with disabilities interact with their environment. It can also address the needs of groups of people or users who experience certain difficulties in using a given product or service and the way in which they can be provided with tools to overcome these barriers. In this regard, young players, elderly users or players who speak no foreign language (especially English) should also be taken into account when designing video games and digital and multimedia products.

Within this framework, translation can promote this type of accessibility, which is linked to social inclusion and intends to bridge the digital divide and achieve a more inclusive society. Providing information and entertainment in several languages should be regarded as a relevant point for accessibility checklists when games are being designed. Users should be able to enjoy video games in their own language in a similar fashion to films or DVDs. However, the limitations imposed by budgetary constraints and technology itself still hamper the multilingual localization of many video games.

Although this paper underlines the importance of a broader concept of accessibility aimed at promoting e-inclusion and allowing access to video games to the widest possible audience, game accessibility currently focuses largely on people with special needs. In this regard, the relationship between translation and accessibility can be explained on the basis of audiovisual translation, as subtitling and closed captioning are two basic tools with which to make games accessible for the deaf and hard of hearing. Here, translation and technology can clearly improve playability for those users who cannot enjoy mainstream titles designed for the general public.

The accessibility of multimedia products is a relevant issue that needs to be further explored by academia. The democratization of the games industry and the introduction of more interested parties and the increase in player numbers in the video games field make this an emerging and exciting area to be explored by researchers, scholars and professionals. In this sector, being multidisciplinary is a prerequisite, since the evolution of video games into multidimensional and multimodal products demands that they are accessible to the largest possible audience. In order to meet this objective and promote knowledge transfer from academia into wider society, the collaboration of researchers with industry professionals is essential in order to make fun available to everyone.

References

Agost, Rosa and Chaume, Frederic (eds). 2001. *La Traducción en los Medios Audiovisuales.* Castellón de la Plana: Universitat Jaume I.

Badia, Toni and Matamala, Anna. 2007. 'La Docencia en Accesibilidad en los Medios' in *Trans* 2: 61–71.

Bernal, Miguel. 2006. 'On the Translation of Video Games' in *The Journal of Specialised Translation* 6: 22–36. Online at: <http://www.jostrans.org/issue06/art_bernal.php> (consulted 18.04.2013).

Bernal, Miguel. 2009. 'Video Games and Children's Books in Translation' in *The Journal of Specialised Translation* 11: 234–247. Online at: <http://www.jostrans.org/issue06/art_bernal.php> (consulted 18.04.2013).

Chandler, Heather. 2005. *The Game Localization Handbook.* Massachusetts: Charles River Media.

Chandler, Heather and Stephanie Deming. 2011. *The Game Localization Handbook.* Massachusetts: Jones and Barlett Publishers.

Díaz-Cintas, Jorge. 1998. 'La Labor Subtituladora en tanto que Instancia de Traducción Subordinada' in *Proceedings of the III Congrés Internacional sobre Traducció:* 83–90. Online at: <http://spiral.imperial.ac.uk/handle/10044/1/1424> (consulted 28.01.2013).

Díaz-Cintas, Jorge. 2001. *La Traducción Audiovisual. El subtitulado.* Salamanca: Almar.

Díaz-Cintas, Jorge. 2003. *Teoría y Práctica de la Subtitulación: Inglés/Español.* Barcelona: Ariel.

Díaz-Cintas, Jorge. 2005. 'Audiovisual Translation Today. A Question of Accessibility for All' in *Translating Today* 4: 3–5.

Díaz-Cintas, Jorge. 2008. *The Didactics of Audiovisual Translation.* Amsterdam: John Benjamins.

Eskelinen, Markku. 2001. 'Towards Computer Game Studies' in *Digital Creativity* 12(3): 175–183.

European Commission. 2006. 'Europeans and their Languages'. *Special Eurobarometre* 243. Online at: http://ec.europa.eu/public_opinion/archives/ebs/ebs_243_en.pdf> (consulted 21.04.2013).

France, Mark. 2007. 'Audio Game Survey Results' in *Audyssey Magazine* 50. Online at: <http://www.accessibility.nl/games/researchdocs/MarkFranceAudioGame2007.pdf> (consulted 15.04.2013).

Frasca, Gonzalo. 1999. 'Ludology Meets Narratology: Similitude and Differences between (Video) Games and Narrative'. Online at: <http://www.ludology.org/articles/ludology.htm> (consulted 28.01.2013).

Frasca, Gonzalo. 2001. *Videogames of the Oppressed: Videogames as a Means for Critical Thinking and Debate.* Master's thesis. Georgia Institute of Technology. Online at: <http://www.ludology.org/articles/thesis/> (consulted 20.04.2013).

Friberg, Johnnyand Gärdenfors, Dan. 2004. 'Audio Games: New Perspectives on Game Audio'. Paper presented at *Advances in Computer Entertainment Technology* (Singapore, 3–5 June 2004). Online at: <http://www.cms.livjm.ac.uk/library/AAA-GAMES-

Conferences/ACM-ACE/ACE2004/FP-18friberg.johnny.audiogames.pdf> (consulted 22.04.2013).

Franco, Eliana, Matamala, Anna and Orero, Pilar. 2010. *Voice-over Translation. An Overview*. Bern: Peter Lang.

Gambier, Yves. 2004. 'La Traduction Audiovisuelle: un Genre en Expansion' in *Meta* 49 (1): 1–11.

Gambier, Yves. 2006. 'Multimodality and Audiovisual Translation'. Paper presented at *MuTra 2006 Audiovisual Translation Scenarios* (Copenhagen, 1–5 May 2006). Online at: <http://www.translationconcepts.org/pdf/2006_Gambier_Yves.pdf> (consulted 21.04.2013).

IGDA. 2004. 'Accessibility in Games: Motivations and Approaches'. Online at: <http://www.igda.org/accessibility> (consulted 12.04.2013).

ISFE. 2010. 'Video Gamers in Europe 2010'. Online at: <http://www.isfe-eu.org/> (consulted 12.04.2013).

Mangiron, Carmen. 2004. 'Localizing Final Fantasy: Bringing Fantasy to Reality' in LISA *Newsletter GlobalInsider* XIII: 1.3.

Mangiron, Carmen. 2010. 'Aplicaciones Educativas de los Videojuegos'. Presentation at *Gamerland* (Bilbao, 31 October 2010).

Mangiron, Carmen. 2011. 'Accesibilidad a los videojuegos: estado actual y perspectivas futuras' in *Trans* 15: 53–67.

Mangiron, Carmen and O'Hagan, Minako. 2006. 'Game Localisation: Unleashing Imagination with Restricted Translation' in *The Journal of Specialised Translation* 6: 10–21. Online at: <http://www.jostrans.org/issue06/art_ohagan.php> (consulted 20.04.2013).

Mayoral, Roberto, Kelly, Dorothy and Gallardo, Natividad. 1988. 'Concept on constrained translation: Non-Linguistic perspectives of translation' in *Meta* 33: 356–367.

Muñoz Sánchez, Pablo. 2008. 'En Torno a la Localización de Videojuegos Clásicos Mediante Técnicas de *Rombacking*: Particularidades, Calidad y Aspectos Legales' in *The Journal of Specialised Translation* 9: 80–95. Online at: <http://www.jostrans.org/issue09/art_munoz_sanchez.pdf> (consulted 20.04.2013).

Neves, Josélia. 2005. *Audiovisual Translation: Subtitling for the Deaf and Hard-of-Hearing*. PhD thesis. London: Roehampton University.

Newman, James. 2004. *Video Games*. New York and London: Routledge.

O'Hagan, Minako. 2005. 'Multidimensional Translation: A Game Plan for Audiovisual Translation in the Age of GILT'. Plenary presentation at *EU High Level Scientific Conference Series: Multidimensional Translation – MuTra* (Saarbrucken, 2–6 May 2005).

O'Hagan, Minako. 2007. 'Video Games as a New Domain for Translation Research: from Translating Text to Translating Experience' in *Tradumática* 5. Online at: <http://www.fti.uab.es/tradumatica/revista/num5/articles/09/09art.htm> (consulted 21.04.2013).

O'Hagan, Minako. 2009. 'Towards a Cross-Cultural Game Design: an Explorative Study in Understanding the Player Experience of a Localised Japanese Video Game' in *The Journal of Specialised Translation* 11: 211–233.

Orero, Pilar. 2005a. 'Audio Description: Professional Recognition, Practice and Standards in Spain' in *Translation Watch Quarterly* 1: 7–18.

Orero, Pilar. 2005b. 'La Inclusión de la Accesibilidad en Comunicación Audiovisual Dentro de los Estudios de Traducción Audiovisual" in *Quaderns: revista de traducció* 12: 173–185.

Orero, Pilar. 2011. 'Audio Description for Children: Once upon a time there was a different audio description for characters' in Di Giovanni, Elena (ed.) *Between Text and Receiver: Accessibility, Dubbing and Translation.* Bern: Peter Lang. 169–184.

Orero, Pilar, Pereira, Ana María and Utray, Francisco. 2007. 'Visión Histórica de la Accesibilidad en los Medios de España' in *Trans* 2: 31–43.

Tavinor, Grant. 2009. *The Art of Videogames.* Oxford: Wiley-Blackwell.

Tercedor, María Isabel. 2005. 'Aspectos Culturales en la Localización de Productos Multimedia' in *Quaderns, revista de traducció* 12: 151–160.

Tercedor, María Isabel. 2010. 'Translating Web Multimodalities: Towards Inclusive Web Localization" in *Tradumática* 8. Online at: <http://www.fti.uab.es/tradumatica/revista/num8/articles/09/9art.htm> (consulted 21.04.2013).

United Nations. 2006. 'Convention on the Rights of Persons with Disabilities and Optional Protocol'. December 16 2006. New York.

Javier Torrente, Ángel del Blanco, Pablo Moreno-Ger, Iván Martínez-Ortiz & Baltasar Fernández-Manjón

Accessible Games and Education: Accessibility Experiences with *eAdventure*

The increasing importance of the video games entertainment industry has prompted different efforts to promote the inclusion of accessibility features within games. Whilst the field of academia has produced relevant and promising results, mainstream producers are still reluctant to invest in access-for-all. However, in the field of academia there is also a growing trend towards introducing games and game-like activities into educational settings – usually labeled as 'serious games' *– in an attempt to explore other learning approaches and to improve the engagement of students. Yet while entertainment games can (arguably) opt to ignore accessibility, educational games* must *be inclusive and cannot afford to ignore it. In this paper, we present our approach to promoting accessible educational games. To reduce the barriers to and costs of creating accessible educational games, we explore the use of game creation tools with built-in accessibility features, as opposed to adding accessibility features* ad hoc *to existing educational games.*

1. Introduction

Along with the increasing demand for improved educational processes, recent years have seen an increase in the application of new technologies and media formats to support new pedagogical strategies in order to prepare students for the challenges demanded by our rapidly changing society. Within the academic field of Technology-Enhanced Learning, there is an emerging trend to provide more dynamic and interactive content for students such as in the form of video games, which have been particularly highlighted because of their potential as learning tools (Gee, 2007; Kirriemuir and McFarlane, 2004; Michael and Chen, 2006).

However, the technology of video games is still an emerging and rapidly growing one. Other technologies – especially the Internet – have also experienced rapid expansion, but these fast-paced advances often entail a marginalization of people with special needs who cannot access digital content. This can be a consequence of a personal disability (e.g. blindness, deafness, reduced mobility, learning disorders, etc.) or even be due to contextual and technological issues (e.g. noisy environments where sound cannot be heard or connectivity is limited, language barriers, etc.).

As a consequence, there has been increasing development of technologies that enhance the accessibility of information systems for people with special needs. Nevertheless, the creation of accessible technologies has been unequally focused on the different fields of software development. Whilst Web accessibility is reasonably well catered for, including multiple initiatives, tools, standards and guidelines, developers involved in other technological areas such as interactive multimedia (and especially video games) are still trying to find the most suitable way to create accessible products (Abrahams, 2010).

Focusing on the specific topic of games, whilst it is true that there are some video games that include accessibility features, the high costs involved in incorporating some of these features in the game post-production is hindering their widespread adoption. Moreover, some of the projects that include accessibility features have been introduced as a result of other motivations (e.g. including in-game voice command support to enhance the gaming experience). Commercial games tend to pay little attention to these accessibility issues, partly because the industry perceives that the return on any investment is marginal. Due to a lack of specific regulations governing accessible commercial games, many publishers opt to ignore the aspect of accessibility.

However, when we focus on educational games, these problems and the discussion relating to the potential return on the investment budget must be examined differently. The need for enhanced accessibility in any kind of educational content is more pressing than in developments that are purely entertainment-driven. According to a recent report on disability jointly produced by the WHO and the World Bank, more than a billion people in the world today experience disability (World Health Organization, 2011). According to this report, the estimated number of children experiencing "moderate or severe disability" ranges from 93 to 150 million, depending on the survey. If educational video games are to play a role in education, accessibility must be considered.

Accessibility for educational video games needs to address a diverse range of issues. Video games provide engaging experiences that are far more complex than other information systems which simply grant users access to data. As such, approaches which bring accessibility to the Web and other information systems are not fully scalable to video games, as these approaches may hinder the games' immersive and engaging atmosphere. In addition, when it comes to the educational field, it is harder to assume the higher cost of accessible games, given that most innovative educational gaming projects often have a limited budget. These contexts require methodologies, design patterns and tools specially devised to facilitate the creation of accessible educational video games. In reality, such elements are rare and have received scant attention in the surrounding literature.

The aim of our work is to facilitate the introduction of accessible features into the development of educational video games without compromising development costs. Integrating accessibility features into the tools used to create the games would free developers from having to implement ad-hoc accessibility solutions for their games. With this objective in mind, we have introduced a set of accessibility features into *eAdventure*, a game-authoring platform designed to facilitate the creation of educational *point-and-click* adventure games.

This paper is structured as follows: Section 2 presents some related work, focusing on the potential issues and current trends in the fields of accessibility, games and education. In section 3 we discuss some design issues especially relevant to educational gaming. Section 4 describes the accessibility features of the *eAdventure* platform. Section 5 presents a case study, in which a pre-existing game is enhanced with accessibility features and the approach evaluated. Finally, section 6 presents some conclusions and potential future lines of research.

2. Related Work

The accessibility of information systems is rapidly attracting the attention of national authorities and IT professionals, since it is one of the potential sources of a digital divide. In this context, the accessibility of educational technologies can seriously affect the future opportunities of individuals who have limited means of access. While traditional teaching methods are often able to cope with aspects of accessibility through the efforts of the instructors, the current trend towards increasingly complex educational technologies is continuously increasing the challenge. In this section, we discuss the relevant topics for state-of-the-art accessibility in Technology-Enhanced Learning in general, and in particular in educational gaming.

2.1 Accessibility in e-Learning Environments

As e-Learning environments are mainly web-based systems (e.g. Learning Management Systems – LMS – such as Moodle™, Blackboard™ or Sakai™), the current state-of-the-art accessibility for e-Learning is very closely related to Web accessibility in general.

E-Learning environments have profited from the ongoing efforts of different public and private organizations to improve WWW accessibility. Highly influential organizations such as the W3C are presenting the necessary requirements for creating accessible web content through the Web Accessibility Initiative (WAI). WAI includes guidelines and techniques for the development and evaluation of multiple types of accessible applications related to the Web (W3C, 2002, 2008, 2011a, 2011b). Along with this initiative, different webmaster-oriented tools have been created which are devoted to checking the level of accessibility of web-based content and applications (W3C, 2006). Besides this, assistive technologies such as screen-readers or screen magnifiers have partially helped to improve the level of accessibility of the Web.

There are also initiatives that specifically deal with digital educational content for web environments. A very thorough approach was undertaken by the IMS Global Consortium in their IMS AccessForAll set of specifications (IMS Global Consortium, 2003, 2004). This initiative tries to define a set of data to describe the needs of students with disabilities and to tag the materials (Learning Objects) accordingly. The content that is delivered to the students could therefore be aligned with their special requirements. A similar approach is the ISO/IEC 24751-1: 2008 standard, developed by the International Organization for Standardization (ISO). Other initiatives focus on the analysis of the level of accessibility of popular e-Learning systems (Freire et al., 2009; Minovic et al., 2008) or on enhancing accessibility of e-Learning systems (Sclater, 2008).

2.2 Special and Adapted Game Devices

The most common approach to increasing the accessibility of video games is to seek their compatibility with assistive technologies (Kearney, 2005). This includes compatibility with adapted and special hardware, but also with software. Some examples are screen-reading tools, mouse emulators and virtual keyboards. There are also tools that can be used to substitute the usual gamepads provided by game consoles (e.g. vocal joysticks, head gamepads or tongue sensors).

Following this line of research, the work presented by Sjöström & Rassmus-Gröhn (1999) shows the use of the PHANToM™ device as an example of how haptic devices (devices which provide human-computer interaction based on body movements and the sense of touch) can increase accessibility. This approach not only facilitates access to the games for a wide range of people with reduced mobility (controlling the video games with easy movements of one finger), but can also be useful for visually impaired people because the device offers them the possibility of perceiving 3D objects by means of the movements and vibrations it produces.

Another approach consists of adjusting games without requiring specific devices (e.g. adding subtitles). However, it is possible to bring both concepts together. For instance, there are games that allow the player to combine screen-reading tools, mouse emulators and virtual keyboards. In the same vein, we find auditory games (also known as "audio-games") (Friberg and Gärdenfors, 2004). These are games specially designed for people with visual disabilities where all the information from the game is transmitted via audio (Röber and Masuch, 2005). Specific sounds with special meanings are used intensively throughout the game so it is easy to remember the association between sounds and their meanings. In some of these games, the indications are given with abstract sounds, but the games most widely accepted are those which give users vocal descriptions reproduced through text-to-speech synthesizers. Other games receive input either vocally or by means of specific devices (Targett and Fernström, 2003).

2.3 Accessibility in Entertainment-Driven Commercial Video Games

There are some commercial video games that implement features to enhance accessibility from development or that have been modified after publication for this purpose. The creators of *Half Life 2*™ (2004) introduced accessibility for people with hearing problems during the development process after they received complaints concerning the first game of the series, *Half Life*™ (1999–). The reason for this is that in *Half Life*™, certain information that was essential in order to complete the game was transmitted through cut-scenes (videos) without subtitles, making it impossible for people with hearing disabilities to reach the end of the game (Bierre et al., 2005).

Another example is *Terraformers*™ (2003), a game that was directly designed to include accessibility features from an early stage. It has a normal mode in which visual graphics are reproduced in the usual manner of first-person 3D games, but it also has an accessible mode. In the latter mode, sonar is activated to tell players what is in front of them and the graphic contrast is increased for visually-impaired people (Westin, 2004). This mode also allows the player to select objects from the inventory using voice commands.

Other academic papers have focused on providing guidelines about how to design interfaces or methodologies for accessible video game development (Friberg and Gärdenfors, 2004; Grammenos, Savidis and Stephanidis, 2007). As yet, there are no broadly accepted standards or specifications in this regard, but there are a few web-based initiatives that provide broad guidelines as to how to develop video games with accessible features. These initiatives must be translated into standards in order to unify the criteria and make the way to create accessible games more clear and facilitate the reuse of successful practices.

The International Game Developers Association (IGDA) has a Special Interest Group which focuses on accessibility issues. This group is active in producing state-of-the-art reports and analysis covering accessibility in games. One of their early works (Bierre et al., 2004) provides a general overview of the field, covering what is meant by accessibility in games, why this is necessary, what kind of disabilities can be tackled at the stage of video game creation, and the most frequent adaptations that developers concerned with accessibility usually perform. The document also outlines how to adapt existing games to improve their accessibility through the addition of subtitles and customizing text fonts, and how the textual information and subtitles can be recorded or synthesized. Along with these ideas, the authors encourage the use of other approaches to gather user input, such as the use of voice recognition or other specific devices. However, the report does not propose any concrete patterns or methodologies for creating accessible games.

From a technological point of view, a unique approach is proposed by FORTH (Foundation for Research and Technology – Hellas), based on the Unified User Interface Design (UUID) (Savidis and Stephanidis, 2004). UUID proposes a design pattern where game tasks are initially considered in an abstract device-independent way. In later design phases, the interaction for each game task is designed, including the selection of input/output devices. Several games have been developed following these guidelines, achieving accessibility for people with a wide range of special needs. These are the universally accessible games (UA-Games). One example is *Access Invaders* (Grammenos, Savidis, Georgalis and Stephanidis, 2006), which supports different game settings depending on the potential disabilities of each player. These include blindness (in which case the game will be loaded with the appropriate characteristics of the Audio-Games), damaged vision and cognitive or motor disabilities.

As far as development tools are concerned, the market is populated with many authoring environments for the development of video games. There are development frameworks for game programming (such as Microsoft™ XNA™), game development environments which allow people without technical knowledge to develop their own video games (such as *Game Maker*™ and *Unity3D*™), and even simple editors oriented to specific game genres (such as *The FPS Creator* and *Adventure Game Studio*). However, none of these initiatives include pre-configured features which target game accessibility or which are oriented to facilitate universal design. Therefore, accessibility has to be implemented from scratch for each individual game and, depending on the flexibility and possibilities for expansion provided by the platform, it may eventually be unfeasible to introduce certain accessibility features (e.g. a text-to-speech engine is not available).

3. Design Strategies for Accessible Educational Video Games

Video games are very different in comparison to other materials (e.g. web-based content), as they pose accessibility barriers that must be thoroughly analyzed. Some of these can be addressed with a slight increase in development cost if they are considered from the beginning, but the investment may grow alarmingly if they have to be implemented *a posteriori*. For instance, a flexible configuration tool for the game parameters (font settings: color, size etc., audio settings, time response gaps and input/output settings) is something "cheap" to implement and effective for the accessibility needs of many common disabilities.

Other perspectives may be the importance of taking into account the compatibility with special or adapted game devices, or the importance of including special tutorials and documentation within the games (Bierre et al., 2005). Many different recommendations regarding the design of video games can be discussed, but in this section we focus on three issues that are especially relevant to educational video games: the choice of an appropriate game genre, the need for fine-grained adaptation support and finally the distribution and deployment of the games.

3.1 Appropriate Genres for Accessible Educational Video Games

Accessibility requirements are very different depending on the game genre. Moreover, not all game genres have the same educational potential.

In order to make them accessible, game experiences must be designed abstractly without committing to any specific device or input/output system. Therefore, where possible, it is better to focus on game genres in which engagement and immersion are obtained thanks to the attractiveness of game tasks, activities and the flow of the game itself, rather than from features such as the game being visually attractive or providing intensive action. Educational games must capture the attention of students and motivate them even when their accessibility features are activated. Otherwise, their positive effects on learning will evaporate.

Point-and-click adventure games, such as the classic *Monkey Island* (1990–) and *Myst* sagas (1993–2005), meet these requirements. This kind of game captures the player's attention by developing an engaging and motivational plot that players uncover as they advance through the game. Elements such as graphics, sounds and special effects are also part of these games, but only as peripheral features to enhance immersion. They promote reflection instead of action, something which is very convenient for people with motor disabilities as it allows them to solve puzzles with no time pressure. As such, *Point-and-click* adventure and story-telling games are particularly appropriate for education (Dickey, 2006).

3.2 Fine-grained Adaptation vs. Coarse-grained Adaptation

The adaptation performed on a game to make it accessible must be fine-grained, that is as finely-tuned to each player as possible. Whilst relying on stereotypes may solve some of the problems, they may exclude some users. If different alternatives may be applicable in the case of a certain student, the optimum option must be always the choice, whilst considering aspects such as which alternative best preserves the engagement and immersion factors of the game or which will make interacting with the game less difficult and/or time-consuming. This approach differs from typical coarse-grained approaches to web-based content which are built on rough categorizations of students according to their disabilities. As opposed to with other kinds of content, within video games it is possible to provide much more finely-tuned adaptive experiences (Houlette, 2004; Magerko, 2008).

3.3 The Distribution and Deployment of Educational Video Games

The processes of delivering, installing and running games must also be accessible. This presents an extra burden in educational settings. Video games usually consume a lot of machine resources and require top-of-the-range computers that are not always present in schools. To tackle this, we could take advantage of current e-Learning systems to ease game delivery and distribution (Torrente, Moreno-Ger, Martínez-Ortiz and Fernández-Manjón, 2009).

Accessing a game that is embedded in a webpage would be easier for students with disabilities, as it does not require any additional setup and they usually have hardware or software aids with which to navigate the Web. This sets a design restriction on the games, as they have to be web-deployable (e.g. using Java technologies or Adobe Flash) and small in size in order that they can be easily distributed via the Internet.

4. The eAdventure Approach

eAdventure is an educational game platform developed by the <e-UCM> research group at the Complutense University of Madrid (Spain) which has been used in the development of different educational games (Moreno-Ger, Blesius, Currier, Sierra and Fernández-Manjón, 2008; Moreno-Ger et al., 2010). The platform is composed of two applications: a game authoring editor (used to create the educational games) and a game engine (used to execute these games). The editor is instructor-oriented and does not require any technical background or programming skills.

Before beginning this work, the *eAdventure* platform already had some features that could facilitate the development of accessible games, especially for e-Learning applications. Firstly, it is focused on the *point-and-click* adventure game genre. Secondly, an audio file can be attached to any text string in a conversation. Thirdly, *eAdventure* includes mentoring mechanisms to help students when they become stuck on puzzles or other challenges. Finally, *eAdventure* allows the configuration of aspects such as the time at which each message is displayed.

In addition, *eAdventure* provides instructors with special features that enhance the educational possibilities of the platform, including mechanisms to track the performance of each user and to adapt the game experience to the needs of different students (Torrente, Moreno-Ger and Fernández-Manjón, 2008). Finally, *eAdventure* games can be deployed via the Web and integrated into a Learning Management System such as Moodle™.

In the following sections we describe the modifications made to the platform to facilitate the development of accessible educational games. The goal of developing this prototype was not to provide a holistic accessibility solution, but rather to investigate the feasibility of implementing accessibility in a game platform directly at the authoring tool level. Multiple simplifications were therefore made, targeting some of the most common disabilities: blindness, deafness, reduced mobility, low vision and some cognitive disabilities.

4.1 Combination of Input/Output Modules

The *eAdventure* platform includes different pre-configured input/output modules to facilitate the inclusion of accessibility in the games. The idea is that game authors should be able to include multi-modal interaction in their titles in order that people with special needs can play them easily by simply using functionalities included in *eAdventure*. In addition, *eAdventure* provides a number of in-game tools that can be included in the games as extra accessibility aids. These modules can be activated/deactivated automatically at the author's discretion.

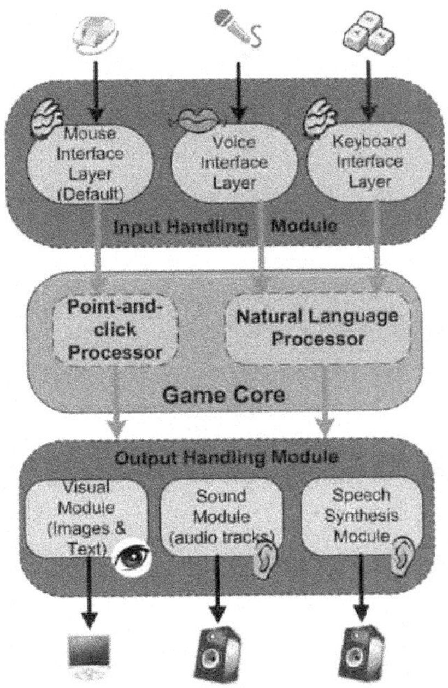

Figure 1: The architecture of the *eAdventure* game engine, with separate layers for input and output.

The *eAdventure* game engine is organized into three layers, with a core layer that handles the game interactions and monitors the state of the game and two separate layers which handle input and output. When the game is launched, it is possible to enable different modules within the input/output layers, selecting the most appropriate combination for the specific needs of each user.

The game core has two separate modules to influence and report the internal state of the game. There is a classic *point-and-click* input processor that was originally available in *eAdventure*. This processor is capable of handling mouse interactions, drawing on-screen graphics and managing simple sound effects. There is also a new Natural Language Processor capable of receiving audio or written input, processing it according to a regular grammar that defines valid commands for the game and using it to modify the state of the game. This module can also produce language output, using pre-recorded audio tracks or a speech synthesis module.

4.2 Input Modules

There are three input modules supported by the *eAdventure* platform: the Mouse Interface Layer, the Vocal Interface Layer and the Keyboard Interface Layer.

The Mouse Interface Layer is the classic interaction mechanism of *point-and-click* adventure games already present in *eAdventure*. Using this interaction mechanism, students usually need to point the mouse over the characters and objects they find as they proceed through the game in order to trigger any kind of in-game interaction. Students therefore need to be able to move the mouse over the screen to discover interactive elements in order to play the games, which may render them inaccessible to students with visual or mobility disabilities.

The Vocal Interface Layer was implemented to allow students with reduced mobility in hands to control games using speech commands. Using a microphone, students can directly "give orders" to trigger any interaction in the game (e.g. "go to the library" or "grab the notebook"). The Keyboard Interface Layer accepts the same orders as the Vocal Interface, but uses the keyboard as the input device. When this layer is activated, a text box appears at the bottom of the screen to allow the user to input commands. Students can thus interact with the game in their natural language, something which can be helpful for students with reduced mobility (using their voice) or visual disabilities (using the keyboard).

Order	*Description*
Examine the table (1)	The game will provide a description of the object "table", if it exists in the scene.
Go to the door (1)	The student's avatar in the game will move towards the place "door".
Grab the pencil (1)	The game will take out the object "pencil" from the scene and put it in the student's inventory[1].
Use ingredient with mixer (1)	The game will combine the objects "ingredient" and "mixer".
Name items in the scene (2)	The game will tell the student which items have already been discovered so that he or she can interact with them.
Open options menu (2)	Pause the game and show the options menu.
Describe (the) scene (2)	The game will provide a description of the scene as a hint for the student.

Table 1: Example of natural language commands available during gameplay. Examples tagged with (1) would be dynamically defined for each scene. Examples tagged with (2) are common to all scenes and games.

Both the Vocal and Keyboard Interfaces were based on the same kind of interaction in order to reduce the implementation cost of the approach and improve the ease of maintenance of the system. Both layers are therefore connected to the same processor (the Natural Language Processor) which receives the commands and maps them onto the game semantic.

The regular grammar that defines game commands combines this kind of dynamically generated rules with some that are constant for all the scenes and games. These rules are used to define basic interactions with the game (e.g. open menus, exit game, skip dialogue lines, etc.). Another important aspect is that in order to enhance usability, the Natural Language Processor accepts diverse synonyms for the verbs and nouns that are fixed (e.g. 'examine the scene' or 'describe the scene' are both permitted). Table 1 shows some examples of typical orders the system would recognize in an *eAdventure* game.

1 The inventory is an element that is usually present in *point-and-click* adventure games. Players use the inventory to store objects they find on their way and keep them for a later use.

4.3 Output Modules

eAdventure has likewise been provided with three output modules: the Visual Output Module, the Sound Output Module and the Speech Synthesis Module.

The Visual Module is not only used to display images on the screen (the background image for the scene, the characters and objects etc.) but also text. Text is a key element in *point-and-click* adventure games, as these games commonly provide information through conversations with other characters which are usually written on screen.

The visual module can also be enhanced with two additional features. Firstly, game authors can provide students with a screen magnifier. To avoid disrupting the immersive atmosphere of the game, this is represented as an object that is included in the student's inventory (Figure 2). The student can use it to turn the mouse pointer into a magnifying glass that can be moved around in the game. Similarly, the player can also activate or deactivate a special high-contrast mode that highlights the interactive objects and partially hides the background (Figure 3).

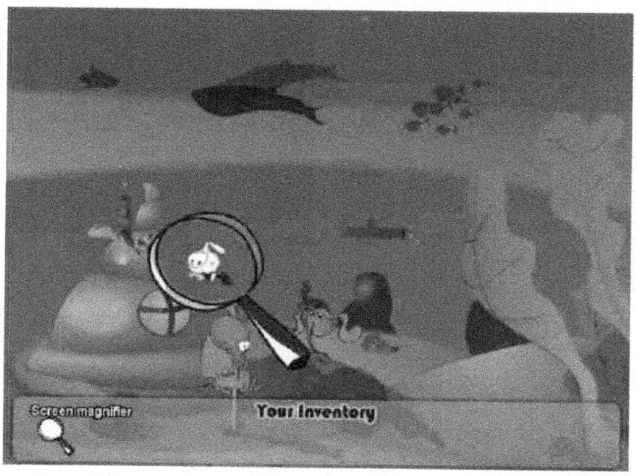

Figure 2: Example of the in-game "screen magnifier" tool in an *eAdventure* game.

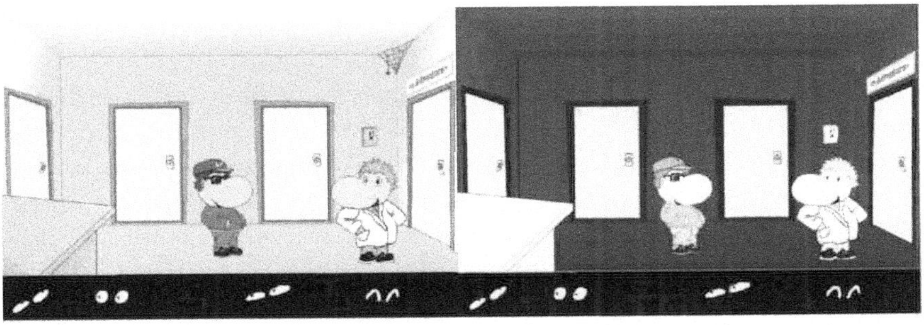

Figure 3: A game scene in regular view (a) and in high-contrast mode (b).

In turn, the Sound Module is capable of playing pre-recorded sounds, and it is possible to use it to add accessibility features to in-game conversations and texts by recording all the dialogues (the sound module can play audio tracks in MP3 format). However, this significantly increases the cost of the game, requiring voice actors to record each individual utterance. This is often a problem when the budget is limited (as is usually the case for most educational projects).

This is where the Speech Synthesis Module can be a helpful addition, as it allows the introduction of accessibility for students with visual disabilities at a low cost. When this module is enabled, any text to be written on the screen will be automatically reproduced through the Speech Synthesis Module. Higher-budget projects can still use the standard sound module (which plays MP3 files) for increased sound quality. In either case, these modules can also read special accessible descriptions that can be attached to each scene in the game.

The regular Sound Output Module is also used to play descriptive sounds as an alternative feedback for the user. For instance, when the Keyboard Interface is activated and the user introduces a command, the system uses special beeps to indicate whether or not the command was a valid one. Analogously, other actions such as entering or leaving the options menu have been associated with other specific beeps.

4.4 Configuring Accessibility Features with the eAdventure *Game Editor*

The first step to create an accessible adventure game is obviously to design and develop the game with the *eAdventure* game editor. It is recommended not to leave the decision about accessibility to the last moment, but to instead think about the accessibility features that are going to be introduced in the game during the design phase. This is especially true if these will require adapting the game flow as this would involve providing alternative paths, dealing with difficulty settings or providing additional aids in some situations.

When the game is designed, the author must select the input/output modules and the in-game accessibility tools (such as the screen magnifier) that will be active in the game. The game editor uses these settings to optimize the exportation process so that no unnecessary modules will be packaged within the game.

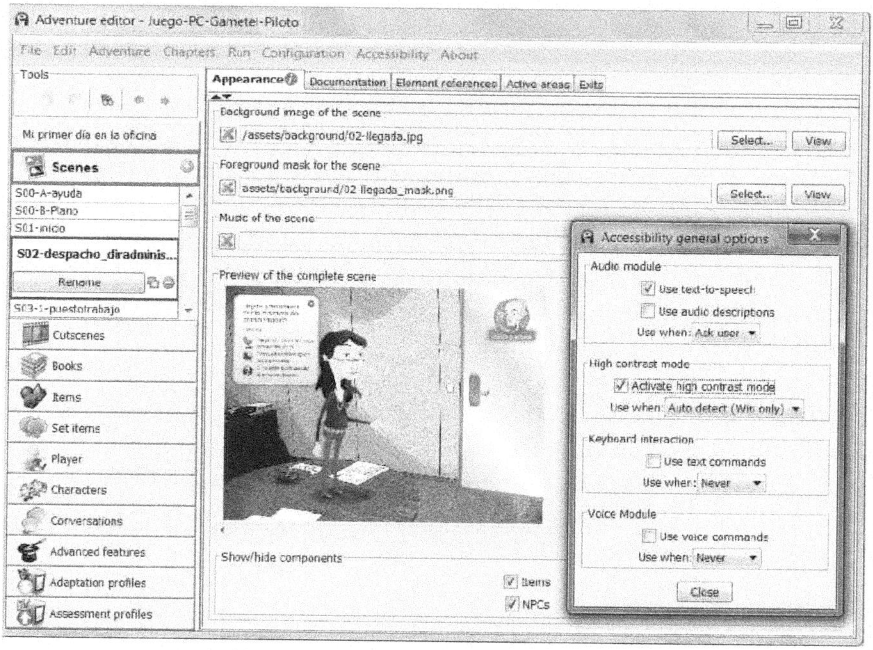

Figure 4: Edition of the Input/Output settings with the *eAdventure* editor.

If visual disabilities are considered, it is very important that all the visual elements of the game receive alternative descriptions. When the player enters a scene, the game engine will use these descriptions along with some extra information that it computes from the game definition (e.g. the number of elements in the scene) to create a complete description of what the student is supposed to see. The completed description is synthesized and played using the audio system. Authors can redefine this behavior by introducing a list of available descriptions for each scene and set the conditions that will trigger each description.

The next step is providing captions for cut-scenes (videos and slides). This feature is currently only supported for slide-scenes in *eAdventure*[2]. Finally, game authors need to create the game adaptation profile which will determine under what circumstances the game must be adapted, and how the adaptation must be carried out.

5. Case Study

As a case study to test the new *eAdventure* accessibility features, we introduced accessibility into a pre-existing game. Following the ideas described in section 4, we introduced accessibility for people with different degrees of visual, auditory, motor and cognitive disabilities in the game *1492*, an educational game about Spanish History[3]. *1492* specifically focuses on the events that occurred in 1492, such as the Granada War and Columbus' expedition (Figure 5). These are notable events in the history of Spain and are thoroughly covered in primary education, which is a strong additional motivation for making the game accessible.

2 A slide-scene is a special type of cut-scene that displays a sequence of static images, rather than full-motion video.

3 The original version of the game can be downloaded from the *eAdventure* game repository: <http://e-adventure.e-ucm.es/course/view.php?id=26>. Video produced by Spanish Public Broadcasting System, available on Youtube: <http://youtu.be/ROg3pjnfi8U>.

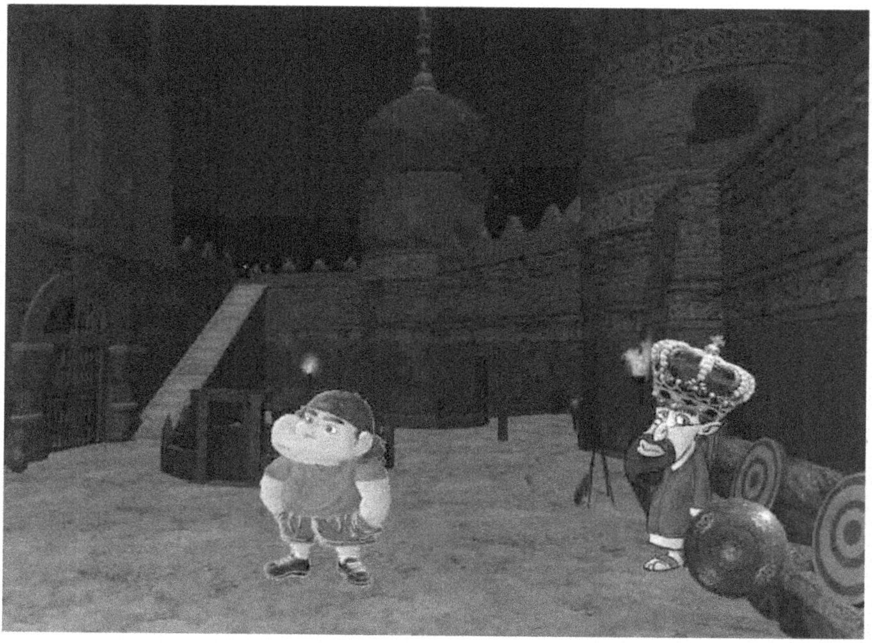

Figure 5: A screenshot from the game *1492*. Cristobalín is exploring King Boabdil's palace looking for the stolen ceremonial Key to the City.

5.1 Adaptation of the 1492 *Game*

1492 was not initially designed as an accessible game. The first step was to decide the target disabilities and to then activate/deactivate the necessary input/output modules and/or in-game tools using the game editor. For this case study we considered visual, auditory, mobility and cognitive problems.

As cognitive disabilities are very complex and may require very different adaptations, we considered just two possibilities in order to test the system: students with low memory capacity and students with non-severe reasoning problems. In the first case we defined alternative conversations that lessened the amount of information that the student is required to gather at any one moment, thus increasing the focus on relevant information and reducing the amount of "superfluous" information. In the second case, we defined alternative game paths with simpler puzzles and riddles. Furthermore, the original *1492* game included an in-game multiple-choice examination at the end of the game by means of a conversation between the main character (a student called Cristobalín) and his teacher. For both types of cognitive disabilities, we provided an alternative, more linear exam.

In order to cover the rest of potential special requirements, the game is configured with all of the input/output modules and the screen magnifier. The high contrast mode was not used. To allow the modules to describe the game, we also had to provide alternative descriptions of the visual elements found in each scene, as well as of the scenes themselves, so that these could be passed to the speech synthesizer.

5.2 Preliminary Evaluation

The preliminary evaluation phase so far conducted involved two end users. In this session, the game was played for 20 minutes by a blind user and a user with reduced mobility in hands. Both users had prior experience interacting with computers both for work and entertainment (they liked to play some accessible video games). For the visually impaired user, the system was configured with the Keyboard Interface and the Speech Synthesis Module activated. For the user with motor disability, the Vocal Interface was activated as the interaction method. During the experience, we observed and documented the reactions of both participants. The most relevant conclusions obtained from this session are as follows:

The blind user had some initial trouble interacting with the game. Apparently he did not find the mechanism for interaction intuitive. He expected to be able to navigate through the game elements using the keyboard arrows and select the interactions from a menu as he would typically do when navigating the Web. After a while, he began managing to interact with the game without making major errors. In this regard, the auditory feedback provided by the system (speech synthesis and special sound effects) seemed to be appropriate. Nevertheless, it was sometimes difficult for the user to identify which character was speaking as not all the characters' voices were different. This person did not need any assistance from the researchers and could complete the game session on his own.

The person with motor disability had some initial problems with the pronunciation of the commands. It was noted by the researchers that the shorter the pronounced commands, the more efficient the voice recognition. Nevertheless, the user did not realize this and became quickly frustrated and he therefore required some help from the researchers to understand how he was expected to interact with the system. Following this, the accuracy of the vocal interface began to increase, allowing him to complete all the three scenes of the game included in the evaluation session plan (in the case of the second user, the full game was not tested in order to keep the session as short as possible). The main issue with the vocal interface was the vocabulary that the player needed to use in order to activate the game commands. As a result of this experiment, we realized that it is necessary to add flexibility to the vocabulary (e.g. introducing more synonyms from a thesaurus) for the different game actions.

6. Conclusions and Future Work

The current trend in learning technologies is towards increasingly complex multimedia and interactive content and this presents a significant accessibility challenge. In this regard, while entertainment-driven games can to some extent afford to ignore accessibility concerns, educational games should be inclusive and available to everyone regardless of their individual conditions.

Nevertheless, the development of accessible games comes at a cost. In educational settings, with both limited budgets and markets, the problem becomes greater. In addition, accessible video games are a relatively new idea and the existing research in the field is still at an early stage. In this work, we have presented the foundations of our approach to accessible educational gaming, which provides a tool to facilitate the inclusion of accessibility features in educational video games.

However, the system is still at the prototype stage. The evaluation thus far performed has proven the feasibility of the approach, as end users were able to interact with the system. Nevertheless, the results obtained show that there are still open issues that should be dealt with before incorporating the features into a production environment. In this regard, according to the results of the evaluation, it would be necessary to reduce the learning curve, as end users become frustrated the first time they interact with eAdventure games using the vocal and keyboard interfaces. This may be solved by including further guidance and an in-game tutorial that explains how the user is expected to interact with the game. Once they are stable enough, we are planning to integrate the accessibility features described throughout this paper into the main release of the open source eAdventure platform for use by the general public. This will probably be when the second generation of the platform (*eAdventure* 2.0) is released, which is initially scheduled for late 2012.

Another aspect that will require further attention is the evaluation of our approach for users with cognitive disabilities. While the case study was designed to cope with some cognitive disabilities, it has yet to be tested with target users. Finally, further testing is required in order to measure whether the introduction of accessibility in the games had a negative impact on the user's immersion and engagement. While the users who tested the system felt positive about the experience, the engagement with the regular version of the game compared to the accessible one has not been formally evaluated.

Acknowledgements

The Spanish Committee of Science and Technology (grant TIN2010-21735-C02-02) have partially supported this work, along with the Complutense University of Madrid and the Regional Government of Madrid (research group 921340 and project e-Madrid S2009/TIC-1650), and the EU through the projects GALA Network of Excellence in Serious Games (FP7-ICT-2009-5-258169), SEGAN (519332-LLP-1-2011-1-PT-KA3-KA3NW) and CHERMUG (519023-LLP-1-2011-1-UK-KA3-KA3MP). Thanks also to our colleagues from Technosite (ONCE Group) for their kind advice and support.

References

Abrahams, Peter. 2010. 'Past Present and Future of ICT Accessibility'. *IT-Analysis.com*. Online at: <http://www.it-analysis.com/business/compliance/content.php?cid=12331> (consulted 16.01.2012).

Bierre, Kevin, et al. 2005. 'Game Not Over: Accessibility Issues in Video Games'. In proceedings at *11th International Conference on Human-Computer Interaction (HCII'05)*. Lawrence Erlbaum Associates, Inc. Online at <http://www.igda.org/sites/default/files/HCII2005_GAC.pdf> (consulted 05.07.2013).

Bierre, Kevin, et al. 2004. *Accessibility in Games: Motivations and Approaches*. Online at: <http://www.igda.org/sites/default/files/IGDA_Accessibility_WhitePaper.pdf> (consulted 16.01.2010).

Dickey, Michele D. 2006. 'Game Design Narrative for Learning: Appropriating Adventure Game Design Narrative Devices and Techniques for the Design of Interactive Learning Environments' in *Educational Technology Research and Development*, 54 (3): 245–263.

Freire, André, et al. 2009. 'Web Accessibility Metrics: Effects of Different Computational Approaches' in Constantine Stephanidis (ed.) *Universal Access in Human-Computer Interaction. Applications and Services*. Vol. 5616: 664–673. Berlin/Heidelberg: Springer.

Friberg, Johnny and Gärdenfors, Dan. 2004. 'Audio games: new perspectives on game audio'. In proceedings at *ACM SIGCHI International Conference on Advances in computer entertainment technology*.New York: ACM. 148–154.

Gee, James Paul. 2007. *Good videogames and good learning: collected essays on video games*. New York: Peter Lang Publishing.

Grammenos, Dimitris, Anthony Savidis and Constantine Stephanidis. 2007. 'Unified Design of Universally Accessible Games' in Stephanidis, Constantine (ed.) *Universal Access in Human-Computer Interaction. Applications and Services*. Vol. 4556. 607–616). Berlin/Heidelberg: Springer.

Grammenos, Dimitris, et al. 2006. 'Access Invaders: Developing a Universally Accessible Action Game' in Klaus Miesenberger, et al. (eds). *Computers Helping People with Special Needs* (Vol. 4061). Berlin/Heidelberg: Springer. 388–395.

Houlette, Ryan. 2004. 'Player Modeling for Adaptive Games' in Rabin, Steve (ed.) *AI Game Programming Wisdom 2*. Boston, MA: Charles River Media. 557–566.

IMS Global Consortium. 2003. 'IMS Learner Information Package Accessibility for LIP, Version 1.0 Final Specification'. Online at: <http://www.imsglobal.org/accessibility/index.html> (consulted 16.01.2012).

IMS Global Consortium. 2004. 'IMS AccessForAll Meta-data, Version 1.0 Final Specification'. Online at: <http://www.imsglobal.org/accessibility/index.html> (consulted 16.01.2012).

Kearney, P. Robert. 2005. 'Playing in the Sandbox: Developing Games for Children with Disabilities'. In proceedings at *DiGRA 2005 Conference: Changing Views – Worlds in Play*. Vancouver, Canada. 1–7.

Kirriemuir, John and McFarlane, Angela. 2004. *Literature review in games and learning*. Bristol: NESTA Futurelab.

Magerko, Brian. 2008. 'Adaptation in Digital' in *Computer* 41 (6): 87–89.

Michael, David and Chen, Sande. 2006. *Serious Games: Games that Educate, Train, and Inform*. Boston, MA: Thomson.

Minovic, Miroslav, et al. 2008. 'Usability Issues of e-Learning Systems: Case-Study for Moodle Learning Management System'. In proceedings at *OTM Workshops '08*. 561–570.

Moreno-Ger, Pablo, et al. 2008. Online Learning and Clinical Procedures: Rapid Development and Effective Deployment of Game-Like Interactive Simulations' in *Lect Notes Comput Sci, Transactions on Edutainment I, 5080*. 288–304.

Moreno-Ger, Pablo, et al. 2010. 'Application of a low-cost web-based simulation to improve students' practical skills in medical education' in *International Journal of Medical Informatics*, 79 (6): 459–467.

Röber, Niklas and Masuch, Maic. 2005. 'Playing Audio-Only Games: A Compendium of Interacting with Virtual, Auditory Worlds'. In proceedings at *DiGRA 2005 Conference: Changing Views – Worlds in Play*. Vancouver, Canada. 1–8.

Savidis, Anthony and Stephanidis, Constantine. 2004. 'Unified user interface design: designing universally accessible interactions' in *Interacting with Computers*, 16 (2): 243–270.

Sclater, Niall. 2008. 'Overcoming Accessibility Issues for eLearners'. Paper presented at *Second Annual International Conference Sunflower: Education Accessibility*. (Technical University of Ostrava), Celadná, Czech Republic.

Sjöström, Calle and Rassmus-Gröhn, Kirsten. 1999. 'The sense of touch provides new computer interaction techniques for disabled people' in *Technology & Disability*, 10 (1): 45–52.

Targett, Sue and Fernström, Mikael. 2003. 'Audio Games: Fun for All? All for Fun?' in Brazil, Eoin and Barbara Shinn-Cunningham (eds). In proceedings at *9th International Conference on Auditory Display*. Boston, MA, USA. 216–219.

Torrente, Javier, Pablo Moreno-Ger and Baltasar Fernández-Manjón. 2008. 'Learning Models for the Integration of Adaptive Educational Games in Virtual Learning Environments'. In proceedings at *Lecture Notes in Computer Science, 5093*. Springer. 463–474.

Torrente, J., et al. 2009. 'Integration and Deployment of Educational Games in e-Learning Environments: The Learning Object Model Meets Educational Gaming' in *Educational Technology & Society*, 12 (4): 359–371.

W3C. 2011a. 'Accessible Rich Internet Applications (WAI-ARIA) 1.0. *W3C Candidate Recommendation*'. on line at: <http://www.w3.org/TR/wai-aria/> (consulted 16.01. 2012).

W3C. 2011b. 'Authoring Tool Accessibility Guidelines (ATAG) 2.0. *W3C Working Draft*'. Online at: <http://www.w3.org/TR/ATAG20/> (consulted 16.01.2012).

W3C. 2008. 'Web Content Accessibility Guidelines (WCAG) 2.0. *W3C Recommendation*'. Online at: <http://www.w3.org/TR/WCAG20/> (consulted 16.01.2012).

W3C. 2006. 'Complete List of Web Accessibility Evaluation Tools'. Online at: from <http://www.w3.org/WAI/RC/tools/complete> (consulted 16.01.2012).

W3C. 2002. 'User Agent Accessibility Guidelines 1.0. *W3C Recommendation*'. Online at: <http://www.w3.org/TR/UAAG10/ >(consulted 16.01.2012).

Westin, Thomas. 2004. 'Game accessibility case study: Terraformers – a real-time 3D graphic game'. In Proceedings at *5th International Conference on Disability, Virtual Reality and Associated Technologies*. Oxford, UK: 95–100.

World Health Organization. 2011. 'World report on disability' in *National Center for Biotechnology Information. U. S. National Library of Medicine* (June 11th 2011). Online at: <http://www.ncbi.nlm.nih.gov/pubmed/21723520> (consulted 16.01.2012).

Javier Mairena

How to Make Universally Accessible Video Games

Considering universal accessibility features from the beginning of videogame development is as important a point to think about as their implementation. This document makes a set of recommendations as to how to make videogames as accessible as possible, and highlights which features a game should include. It also provides a classification of the main disability user groups the needs of whom should be taken into consideration, and proposes solutions for each of these.

1. Introduction

This document has taken as its reference thirty-eight functionality classifications included in the World Health Organization's 'International Classification of Functioning, Disability and Health' (ICF) that may cause problems during gameplay. This system classifies these functionalities into four groups: cognitive, mobility-related, visual and auditory. (See appendix B)

Two other functionality groups were not included because nowadays, people in these groups have few difficulties related to game accessibility. An easy solution is the creation of an alternative way of controlling these games. The discarded functionalities are voice and speech, and vestibular functions (associated with sensory functions of the inner ear relating to position, balance and movement). A game will be accessible for these functionalities if it offers a method of control that doesn't depend exclusively on balance or voice commands.

2. The Importance of a List of Solutions

To develop videogames and achieve the greatest possible accessiblity, it is important to have a list of accessibility solutions that can be reviewed at the outset and during development, so that developers always consider implementing the maximum number of solutions in a game. Although there are still not many instances of videogame accessibility, solutions already exist that have been applied in existing games. In addition, we can consider more solutions through the problems that currently arise when playing these current games. To have one list, although basic and still based on insufficient experience, will greatly aid the development of accessibility features in the game.

2.1 Pre-existing Lists

Some lists compiling accessibility solutions for video games already exist. Here are some of most interesting:

'Wish List for Accessible Game Design'. Online at: <http://www.gamebase.info/magazine/read/wish-list-for-accessible-game-design_531.html> (consulted 16.07.2011).
'IGDA GASIG Top 10 list'. Online at: <http://gameaccessibility.blogspot.com/2010/11/gasig-top-10-ways-to-improve-game.html> (consulted 16.07.2011).
'Guidelines for Building Blind-Accessible Computer Games'. Online at: <http://www.blindcomputergames.com/guidelines/guidelines.html> (consulted 16.07.2011).
'Guidelines for developing accessible games'. Online at: <http://gameaccess.medialt.no/guide.php> (consulted 16.07.2011).
'Developing For Different Types of Disabilities'. Online at: <http://www.brannonz.com/accessibility/disabilities.html> (consulted 16.07.2011).

3. The Importance of some kind of Structure

While the lists of solutions mentioned in the previous section are very valuable, one thing is missing: a ranked order of the importance of each solution.

As with any other type of software, in the development of video games it is very important to prioritize the functionalities of the product, particularly if using the SCRUM methodology. A single list prioritizing the solutions will help to ensure that the project will at least include the most important and relevant solutions, even though it cannot include them all.

Currently, solutions are initially included according to relevance, with new solutions added and improvements to existing solutions made with each update. The purpose of this paper is to provide a single list setting out all possible solutions according to their importance, thus guaranteeing an improvement in the accessibility of the final product.

4. Solutions and their Classification

Before embarking on the study, forty-five possible solutions to increase accessibility were listed in no particular order. (See appendix A).

A table showing useful solutions for each functionality was created for each group (cognitive, mobility-related, visual and auditory). These tables point out how many times each solution has been marked as a good way to solve a problem. (See appendix C). In this way, a list of solutions sorted by their relevance to each group was created.

Within each list there are many solutions with the same score. These have therefore been divided into groups based on their score. The first group of solutions received 9 points, the second 8 points, and so on. These scores were added in order to create a general list of solutions sorted by relevance.

For clarification, to emphasize a solution relating to its importance for a functionality, a full scientific study was not performed. Instead, the conclusions are based only on general knowledge of the difficulties encountered during gameplay for people with problems relating to these functionalities. Even so, this list provides an initial insight into the important need to prioritize appropriate solutions when developing video games.

5. Lists of Solutions by Functionality

In the following lists, solutions have been sorted by relevance. Groups of solutions separated by line breaks contain those which obtained the same score, and are thus ranked equally.

5.1 Cognitive

- The possibility of going backwards and forwards in the game.
- A guide/summary of what has previously happened in the game.
- Clear objectives throughout the game.
- Clear indicators to direct the player to the next objective.
- Adjustable game speed.
- Simple language and vocabulary.
- An autosave function.
- Different modes of difficulty.
- A simplified graphics mode.
- The ability to save and load the game at any time.
- No time limits in the game.
- The concepts and objectives of the game are easy to understand.
- Sound reinforcement following the achievement of an objective.
- Visual reinforcement following the achievement of an objective.
- A mode involving simplified objectives.
- In-game text spoken out loud.
- In-game text spoken out loud in several languages.
- Icons or pictures to support any text.
- The inclusion of training levels.
- Indicators to determine the mood of the characters in the game (sad, happy, tired etc.) and their attitude to the player (e.g. angry or happy, hate/love him etc.).
- Clear indicators so that the player always knows whether something that happens in the game is good or bad for the player, and which elements and characters in the game are good and bad.
- Different shapes for each kind of element (i.e. anything visible to the player in the game).
- An invulnerability mode.

- No set time limit in which to read text.
- The option to pause gameplay at any time.
- No requirement to simultaneously press more than one button.
- No need to hold down a button in order to perform an action.
- Vibration of the controller to reinforce events in the game.
- Subtitling of all dialogues.
- The possibility to customise controls.
- Control modes involving a reduced number of buttons (included in the switch mode).
- One-handed control mode.
- Configurability of control sensitivity.
- Microphone control mode.

5.2 Mobility

- The possibility going backwards and forwards in the game
- Different modes of difficulty.
- An invulnerability mode.
- Adjustable game speed.
- No need to hold down a button in order to perform an action.
- No need for high-speed repeated keystrokes.
- Configurability of control sensitivity.
- Microphone control mode.
- Head-tracking control mode.
- No time limits in the game.
- The possibility to customise controls.
- No requirement to simultaneously press more than one button.
- Control modes involving a reduced number of buttons (included in the switch mode).
- One-handed control mode.
- The option to pause gameplay at any time.
- The inclusion of training levels.
- An autosave function.

5.3 Visual

- Spoken texts.
- No set time limit in which to read text.
- Dialogue text spoken in several languages.
- A high contrast mode.
- Sound reinforcement following the achievement of an objective.
- Vibration of the controller to reinforce events in the game.
- Adjustable brightness and contrast settings.
- Sounds to act as indicators to play the game without being able to see.
- The option to pause gameplay at any time.
- Different shapes for each kind of element.
- An autosave function.
- The possibility going backwards and forwards in the game.
- A simplified graphics mode.
- Text that can be easily read (visually).
- The option to resize graphical elements.
- Different modes of difficulty.
- An invulnerability mode.
- Icons or pictures to support any text.
- The ability to save and load the game at any time.
- Clear indicators to direct the player to the next objective.
- No time limits in the game.
- Adjustable game speed.

5.4 Auditory

- Subtitling of all dialogues.
- Independent volume settings.
- Use of sign language.
- Closed captions.
- Visual reinforcement following the achievement of an objective
- Vibration of the controller to reinforce events in the game.
- Closed caption when a sound is emitted by an element, or an indication as to where a sound came from if source is not visible on-screen.

6. List of General Solutions

This list is the result of combining the scores for each functionality on the previous four lists. Solutions here are therefore sorted by relevance, but in this instance are not arranged according to groups of functionality. As in the previous lists, there is a line break to differentiate groups of solutions equally ranked due to having obtained the same score.

- The possibility of going backwards and forwards in the game.
- Different modes of difficulty.
- Adjustable game speed.
- Vibration of the controller to reinforce events in the game.
- An autosave function.
- The option to pause gameplay at any time.
- No time limits in the game.
- An invulnerability mode.
- Sound reinforcement following the achievement of an objective.
- Spoken texts.
- Dialogue text spoken in several languages.
- Visual reinforcement following the achievement of an objective.
- A simplified graphics mode.
- No need to hold down a button to perform an action.
- No set time limit in which to read text.
- Clear indicators to direct the player to the next objective.
- Configurability of control sensitivity.
- Microphone control mode.
- No requirement to simultaneously press more than one button.
- Different shapes for each kind of element (i.e. anything visible to the player) in the game.
- Subtitling of all dialogues.
- A guide/summary of what has previously happened in the game.
- Clear objectives throughout the game.
- No need for high-speed repeated keystrokes.
- Head tracking control mode.
- The possibility to customise controls.
- High contrast mode.
- Adjustable brightness and contrast settings.

- Sounds to act as indicators to play the game without being able to see.
- Independent volumes settings.
- Use of sign language.
- Closed captions.
- Control modes involving a reduced number of buttons (included in the switch mode).
- One-handed control mode.
- The inclusion of training levels
- The ability to save and load the game at any time.
- Closed caption on sound emitting element, or indicate where sound came from if the emitter is not visible on-screen.
- Use of simple language and vocabulary.
- Icons or pictures to support any text.
- The concepts and objectives of the game are easy to understand.
- Text that can be easily read (visually).
- The option to resize graphical elements.
- A mode involving simplified objectives.
- Indicators to determine the mood of the characters in the game (sad, happy, tired etc.) and their attitude to the player (e.g. angry or happy, hate/love him etc.).
- Clear indicators so that the player always knows whether something that happens in the game is good or bad for the player, and which elements and characters in the game are good and bad.

7. Making a Universally Accessible Video Game

The most important thing to think about in creating an accessible game is accessibility from the beginning of the game design process. This is because only slight changes in game design could make the implementation of accessibility solutions possible.

One must also bear in mind that the method of creating an accessible game is dependent on the type of videogame. Some solutions are too expensive to develop for certain type of games, meaning that often the developer would be forced to make major changes to the game design in order to implement a particular solution. For this reason, possible development scenarios could be divided into two main categories:

If a developer wanted to create an accessible game to cater to a specific function group, they could consider the main list of accessibility solutions for this particular group whilst working on the game design. In this case, they should be willing to change a large number of aspects of the original game in order to facilitate the implementation of solutions. A better approach would perhaps be to think about game design from the beginning, and implement main accessibility solutions from the start.

In the case where the developer has to make a more accessible a game but without the involvement of big design changes, or when they are not working for a particular function group, it would be a good idea to begin implementing those solutions at the top of the list sorted by their general relevance. Assessing the costs of implementation and taking into consideration the design changes that will be needed will help the developer decide whether or not the particular solution is worthwhile. If not, they should then think about exchanging this solution for the next on the list. It is important to remember that these changes must be made from the first phase of game development.

References

'Developing For Different Types of Disabilities'. Online at: <http://www.brannonz.com/accessibility/disabilities.html> (consulted 16.07.2011).

'Guidelines for Building Blind-Accessible Computer Games'. Online at: <http://www.blindcomputergames.com/guidelines/guidelines.html> (consulted 16.07.2011).

'Guidelines for developing accessible games'. Online at: <http://gameaccess.medialt.no/guide.php> (consulted 16.07.2011).

'IGDA GASIG Top 10 list'. Online at: <http://gameaccessibility.blogspot.com/2010/11/gasig-top-10-ways-to-improve-game.html> (consulted 16.07.2011).

'International Classification of Functioning, Disability and Health (ICF) of World Health Organization'. Online at: <http://apps.who.int/classifications/icfbrowser/> (consulted 01.11.2011).

'Wish List for Accessible Game Design'. Online at: <http://www.gamebase.info/magazine/read/wish-list-for-accessible-game-design_531.html> (consulted 16.07.2011).

Appendix I – Accessibility Solutions

The following accessibility solutions have been listed in no particular order:

1 – An autosave function.
2 – Possibility going backwards and forward in game.
3 – A guide/summary of what has happened previously in the game.
4 – Clear objectives throughout the game.
5 – The ability to save and load the game at any time. Save and load game at any time
6 – Clear indicators to direct the player to the next objective.
7 – Different modes of difficulty.
8 – An invulnerability mode.
9 – No time limits in the game.
10 – A simplified graphics mode.
11 – A mode involving simplified objectives.
12 – Adjustable game speed.
13 – Icons or pictures to support any text.
14 – The inclusion of training levels.
15 – In-game text spoken out loud.
16 – No set time limit in which to read text.
17 – Simple language and vocabulary.
18 – The option to pause gameplay at any time.
19 – The concepts and objectives are easy to understand.
20 – Subtitling all dialogues.
21 – Closed captions.
22 – Independent volumes settings.
23 – Text that can be easily read (visually).
24 – Dialogue spoken in several languages.
25 – High contrast mode.
26 – The option to resize graphical elements.
27 – Enough sounds to act as indicators to play the game without being able to see.
28 – Sound reinforcement following the achievement of an objective.
29 – Visual reinforcement following the achievement of an objective.

30 – The possibility of customising controls.

31 – Control modes involving a reduced number of buttons (included in the switch mode).

32 – One-handed control mode.

33 – No requirement to simultaneously press more than one button.

34 – No need to hold down a button in order to perform an action.

35 – No need for high-speed repeated keystrokes.

36 – Configurability of control sensitivity.

37 – Microphone control mode.

38 – Indicators to determine the mood of the characters in the game (sad, happy, tired etc.) and their attitude to the player (e. g. angry or happy, hate/love him etc.).

39 – Vibration of the controller to reinforce events in the game.

40 – Clear indicators so that the player always knows whether something that happens in the game is good or bad for the player, and which elements and characters in the game are good and bad.

41 – Head-tracking control mode.

42 – Different shapes for each kind of element.

43 – Adjustable brightness and contrast settings.

44 – Closed caption when a sound is emitted by an element, or an indication as to where a sound came from its source is not visible on-screen.

45 – Use of sign language.

Appendix II – Functional Classifications

For this paper, functional classifications are taken from the 'International Classification of Functioning, Disability and Health (ICF)' of The World Health Organization

Cognitive Impairment

b110 Problems with **consciousness functions** (*continuity and quality of consciousness; loss of consciousness* etc.).

b114 Orientation functions (functions of orientation relating to time, place and person).

b117 Intellectual functions (functions of intellectual growth; intellectual retardation, mental retardation etc.).

b122 Global psychosocial functions (interpersonal skills needed to establish reciprocal social interactions).

b126 Temperament and personality functions (extraversion, introversion, agreeableness, conscientiousness, psychic and emotional stability, and openness to experiences; optimism; novelty seeking; confidence; trustworthiness).

b130 Energy and drive functions (functions of energy level, motivation, impulse control etc.).

b134 Sleep functions (the amount and quality of sleep; insomnia, hypersomnia and narcolepsy etc.).

b140 Attention functions (sustaining attention, shifting attention, dividing attention, sharing attention; concentration; distractibility).

b144 Memory functions (short-term and long-term memory; immediate, recent and remote memory; remembering etc.).

b152 Emotional functions (appropriateness of emotion, regulation and range of emotion; ability to affect emotion; sadness, happiness, love, fear, anger, hate, tension, anxiety, joy, sorrow; lability of emotion; flattening of affect).

b156 Perceptual functions (Mental functions of recognizing and interpreting sensory stimuli. Auditory, visual, olfactory, gustatory, tactile and visuospatial perception, such as a hallucination or illusion).

b160 Thought functions (pace, form, control and content of thought).

b164 Higher-level cognitive functions (decision-making, abstraction and organization of ideas; time management, insight and judgement; concept formation, categorization and cognitive flexibility).

b167 Mental functions of language (functions of the expression of spoken, written or other forms of language).

b172 Calculation functions (determination, approximation and manipulation of mathematical symbols and processes).

b180 Experience of self and time functions (awareness of one's identity, one's body, one's position in the reality of one's environment and of time).

Visual Impairment

b2100 Visual acuity functions (sensing form and contour – both binocular and monocular – for both distant and near vision).

b2101 Visual field functions (the entire area that can be seen with fixation of the gaze – related problems include scotoma, tunnel vision and anopsia).

b21020 Light sensitivity and adaptation to the dark – impairments such as night blindness (hyposensitivity to light) and photophobia (hypersensitivity to light).

b21021 Colour vision (problems differentiating and matching colours).

b21022 Contrast sensitivity (separating figures from the ground, involving the minimum amount of luminance required).

b21023 Visual picture quality (seeing stray lights, affected picture quality (floaters or webbing), picture distortion, and seeing stars or flashes).

b2150 Functions of internal muscles of the eye (muscles inside the eye, such as the iris, that adjust the shape and size of the pupil and lens).

b2151 Functions of the eyelid (such as the protective reflex).

b2152 Functions of the external muscles of the eye (the ability to look in different directions, to follow an object as it moves across the visual field, to produce saccadic jumps in order to catch up with a moving target, and to fix the eye).

b2153 Functions of the lachrymal glands (tear glands and ducts).

b220 Sensations associated with the eye and its adjoining structures. (Sensations and related feelings of tired, dry and itching eyes).

Auditory Impairment

b2300 Sound detection (sensing the presence of sounds).

b2301 Sound discrimination (differentiation of ground and binaural synthesis, separation and blending).

b2302 Localisation of sound source (determining the location of the source of sound).

b2303 Lateralization of sound (determining whether the sound is coming from the right or left side).

b2304 Speech discrimination (determining spoken language and distinguishing it from other sounds).

Mobility Impairment

b147 Psychomotor functions (psychomotor retardation, excitement and agitation, posturing, etc).

b176 The mental function of sequencing complex movements (sequencing and coordinating complex, purposeful movements).

b710–b729	Functions of the joints and bones (range and ease of movement of a joint; maintenance of the structural integrity of the joints; range and ease of movement of the scapula, pelvis, carpal and tarsal bones.
b730–b749	Muscle functions (power functions, tone functions, endurance functions)
b760	Control of voluntary movement functions (control over and coordination of voluntary movements)
b765	Involuntary movement functions (unintentional, non- or semi-purposive involuntary contractions of a muscle or group of muscles).

Appendix III – Tables of Solutions and Functionalities

Solutions

Functionalities

Cognitive	1	2	3	4	5	6	7	8	9	10	11	12	13	14	15	16	17	18	19	20	21	22	23	24	25	26	27	28	29	30	31	32	33	34	35	36	37	38	39	40	41	42	43	44	45
b110	X	X	X	X	X																																								
b114			X	X		X																																							
b117	X	X	X	X	X	X	X	X	X	X	X	X	X	X	X	X	X	X	X					X				X	X	X	X	X	X	X				X	X						
b122		X	X	X							X				X		X		X					X														X							
b126							X	X	X			X		X																								X							
b130	X	X					X		X			X																X	X				X	X	X	X									
b134	X	X	X	X	X																																								
b140						X				X	X	X	X		X		X		X					X				X	X										X						
b144	X	X	X	X	X	X				X																																X			
b152																	X											X	X											X					
b156			X	X		X				X							X																									X			
b160	X	X	X	X	X	X					X	X					X		X									X	X											X					
b164		X	X	X		X	X		X	X		X		X				X																						X		X			
b167													X		X	X	X			X				X																					
b172		X					X																																						
b180		X	X	X		X	X		X	X		X							X																										
	6	10	10	10	5	8	6	2	5	6	4	7	3	3	4	2	7	2	5	1				4				5	5	1	1	1	2	2	1	1		3	2	3		3			

Solutions

Functionalities

Mobility	1	2	3	4	5	6	7	8	9	10	11	12	13	14	15	16	17	18	19	20	21	22	23	24	25	26	27	28	29	30	31	32	33	34	35	36	37	38	39	40	41	42	43	44	45
b147																																													
b176																																													
b710-b729																																													
b730-b749																																													
b760																																													
b765																																													
	1	6					6	6	5			6		1				2												5	4	4	5	6	6	6	6				6				

Solutions

Functionalities

Visual	1	2	3	4	5	6	7	8	9	10	11	12	13	14	15	16	17	18	19	20	21	22	23	24	25	26	27	28	29	30	31	32	33	34	35	36	37	38	39	40	41	42	43	44	45
b2100																																													
b2101																																													
b21020																																													
b21021																																													
b21022																																													
b21023																																													
b2150																																													
b2151																																													
b2152																																													
b2153																																													
b220																																													
	5	5			2	2	3	3	2	5		2	3		7	7		6					4	7	7	4	7	7											7			6	7		

Solutions

Functionalities

Hearing	1	2	3	4	5	6	7	8	9	10	11	12	13	14	15	16	17	18	19	20	21	22	23	24	25	26	27	28	29	30	31	32	33	34	35	36	37	38	39	40	41	42	43	44	45
b2300																																													
b2301																																													
b2302																																													
b2303																																													
b2304																																													
																				3	2	3							2										2					2	3

Section II: Game translation

Ornella Lepre

Divided by Language, United by Gameplay: An Example of Ludological Game Localization

Making video games accessible to players from different countries is a complex process that must work across several dimensions to be successful. Based on a game released in different versions in Japan and Western countries, this paper proposes an interpretation of remakes as an extreme form of localization and analyzes the various degrees to which a game can be adapted for a foreign audience. Focusing on the cultural aspects, examples are provided in order to stress how the peculiar nature of games affects textual and audiovisual content as well as the approach to their adaptation. This paper argues that, despite the similarities games share with other media, their localization differs in its unique connection with gameplay, a factor which emerges as a cardinal element for the entire process.

1. Introduction

The video game industry has rapidly evolved and expanded over the last thirty years. According to recent data (ESA, 2010), two-thirds of American homes own either a console or a PC used for gaming. The players are of all ages and both genders and, in many countries, video game sales have surpassed those of the music and movie industries, generating revenues of more than $40 billion worldwide (ibid.). However, the popularity of the medium is not as yet reflected in a comparable academic interest. Game studies is now a recognized academic field but there is little research into games from the perspective of translation studies, despite the importance of localization to the global success and growth of the industry.

While some of the existing literature covers the technical aspects of localization (see Bernal-Merino, 2007, on the methodology of video games translation, or Muñoz Sánchez, 2009, on the translation carried out by fans modifying games through the method known as 'ROM hacking'), a number of works center on its linguistic features. In the debate on the nature of localization, the term *transcreation* introduced by Mangiron and O'Hagan (2006) emerges as a way of denoting the level of freedom translators have in this process. The authors point out that, although creativity is not exclusive to game translation, it permeates it, as what matters is not loyalty to the original text but "to the overall game experience" (ibid.: 15). This view parallels the *ludological* approach proposed in game studies (Frasca, 1999), which views games as a new formal system, with specific rules defined by the dimension of play. This differs from the *narratological* approach, in which games are simply a new form of narrative. O'Hagan (2007) highlights the way that ludological perspectives can affect localization since the experience provided by the ludic nature of games is what localization should aim to translate.

On the practical side, game translation suffers from a lack of universal standards. The difficulty, stressed by Bernal-Merino (2007) and Mangiron and O'Hagan (2006), of achieving a standardization that encompasses all games stems from the varied nature of game content and textual types across the different genres.

For the purposes of this paper, it is worth mentioning those studies that focus on the cultural issues that arise in adapting games for foreign markets. Mangiron and O'Hagan (2006) provide examples from the popular *Final Fantasy* series (1987–) to show how adding dialects and idiomatic expressions, or even new jokes and cultural references, can bring games closer to their target audience. Mangiron (2010) expands on this analysis, paying special attention to the translation of humor by providing examples of puns, plays on words and humorous expressions taken from the *Final Fantasy* series, and points out that a distinctive trait of humor in games is how it is adapted to the culture of the target market. In a similar vein, Di Marco (2007) looks at examples of cultural adaptation in both the textual and visual elements. While stressing the need for such modifications when the original content would be too distant from the target culture or would not conform to acceptability standards (see also Dietz, 2006), she questions the desirability of extreme domestication that might result in the loss of the original identity of the product. Mandiberg (2009) also criticizes the domesticating power of localization as opposed to translation. Comparing several versions of the game *Kingdom Hearts* (2002) created for the American and Japanese markets, he argues that some games should be seen not only as entertainment, but also as political texts, and that localization can remove such undertones.

Considering the above factors, this paper aims to analyze some issues relating to localization, with special attention to the cultural aspects of the process, by using as case studies the game *Elite Beat Agents* (2006) and its Japanese original, *Osu! Tatakae! Ouendan!* (2005). Whilst sharing common traits with both comics and movies, these games display several features unique to the medium, both in the type of content that must be translated and in the manner in which it is presented to the user.

Section 2 introduces the *rhythm games* genre and the two games chosen for the analysis. Section 3 illustrates examples of the localization of *EBA* from English into Italian, drawing comparisons with other media and focusing on the treatment of culture-specific items. Section 4 explains the main differences between the Japanese *Ouendan* and the westernized *EBA*, showing how localization drastically modified the original game while leaving the core experience unchanged, something that was possible because gameplay is the defining component of the two titles. Section 5 then concludes.

2. Playing with Rhythm: *Elite Beat Agents* and *Osu! Tatakae! Ouendan!*

Elite Beat Agents (EBA) and *Osu! Tatakae! Ouendan! (Ouendan)* are *rhythm games* – a particular genre of action game that challenges the player's sense of rhythm. Players must press the correct button at the correct time or, in more general terms, perform the correct action at the correct time (Rollings and Adams, 2006). Although rhythm games originated with earlier titles, their sudden rise in popularity began with the 1996 Playstation game *PaRappa the Rapper*. Rhythm games have proliferated in both Japan and in the West, and the genre has evolved along with the peripherals designed to play them. These peripherals, which are not usually necessary to play the games but greatly enhance the experience, range from dance platforms (*Dance Dance Revolution*, 1996) to maraca-shaped motion sensor controllers (*Samba de Amigo*, 1999), to more complex instrument-shaped input devices. The *Guitar Hero* (2005–) and *Rock Band* (2007–) franchises are arguably the most popular examples of these games, with guitars, keyboards and drum sets that increasingly double as real musical instruments.

While this paper focuses on solutions adopted to remove linguistic and cultural barriers, it is interesting to note that rhythm games have been studied as a possible route to increasing game accessibility. Yuan and Folmer (2008) argued that, since rhythm games rely mainly on audio signals, they are particularly suitable for people with visual impairments. That said, these titles usually require the player to respond to both audio and visual cues. To overcome this difficulty, the authors adapted *Guitar Hero* for blind gamers, substituting visual stimuli with haptic stimuli transmitted to the player through a special glove. The resulting game, called *Blind Hero*, demonstrates how the accessibility of certain mainstream titles can be enhanced without completely altering the experience. In a similar project, Miller, Parecki and Douglas (2007) developed *Finger Dance*, an original rhythm game based exclusively on audio stimuli, and tested it with sighted and visually impaired players. Their results showed that blind or partially sighted people could enjoy the game and play it on their own.

There are a number of reasons for the choice of *EBA* and *Ouendan*. Firstly, they are both games in which the narrative line does not play an important role. In genres that assign a lot of weight to the story, games are becoming increasingly like movies, at least in terms of plot and narrative style. In popular titles, such as *Uncharted: Drake's Fortune* (2007), *Mass Effect* (2007) or the *Final Fantasy* series, long dialogues (often dubbed and subtitled) and cinematic scenes have much in common with their movie counterparts; translators can thus adapt a set of existing resources and techniques to the game medium. *EBA* and *Ouendan*, despite borrowing stylistic conventions from comics, are strictly defined by their ludic component (what is usually referred to as *gameplay*), something that is reflected in their translation. Both games feature many culture-specific items that pose a challenge to translators, who face a trade-off between loyalty to the original and accessibility. In games, this trade-off is not solely limited to the text, but involves all aspects of the product. The final reason for choosing *EBA*, which is the version of *Ouendan* released for the Western market, is that it provides a good example of all-encompassing localization.

Understanding how *EBA* and *Ouendan* work requires a brief discussion of the main characteristics of the Nintendo DS console. The DS is a handheld console with two screens – the lower display is a touchscreen that allows a player to input commands by touching the display with fingers or the provided stylus, in addition to using the standard buttons. This removes one of the layers between the player and the machine, enabling a different, deeper interaction. In some games, the player can interact with specific elements on the screen simply by touching them, by dragging them, or by writing or drawing on the display. The touchscreen makes the DS an ideal platform for rhythm games, which require the player to perform actions such as tapping the screen or drawing specific shapes in rhythm. *EBA* and *Ouendan* are two of the most popular rhythm games to be found amongst the *DS* titles.

3. *Elite Beat Agents*

iNiS, a Japanese games developer specializing in rhythm games, developed *EBA* for the *Nintendo DS* handheld console. Nintendo released the North American version in November 2006, followed by the Australian version in May 2007, and localized versions for the French, Italian, German and Spanish (FIGS) markets in July 2007. The game was translated for the Korean market and released in November 2007.

The bizarre premise of *Elite Beat Agents* is that three special agents try to help people around the world with everyday problems. They accomplish this by dancing to a soundtrack of pop tunes. If they can keep up with the complicated dance routines, the missions are successful; if they fail to follow the steps, the game ends. The upper *DS* screen displays the stories of the different characters in need of help using comic book-style animated panels that evolve towards a better or worse ending according to the player's performance. The action takes place on the lower screen.

As the music progresses, small numbered circles appear on the lower display. The first circle is marked 1, the second 2, and so on. The player must tap the circles with a stylus in the correct order and in time with the music. He or she can identify when to tap each circle by following the rhythm of the song, but the game also provides visual clues. A large circle surrounding each small numbered circle gradually shrinks to the size of the small one until the moment to tap it arrives.

Occasionally, the player must perform other actions such as sliding the stylus from one circle to another or rapidly spinning it on the screen. The success in each stage is determined by a health meter that goes up or down depending on whether the player performs the right actions. If the meter reaches zero before the end of the song, the mission fails.

While the majority of rhythm games do not display a significant amount of text (this largely comprises the menus, documentation and packaging), the accompanying stories in *EBA* required more effort to localize. Most of the text is embedded in graphics, which can introduce an additional complication and impose further space constraints on the translation.

3.1 Rhythm localized – a text analysis

This section compares the English source text in representative stages of the game with its Italian translation. The specific levels are chosen based on the background stories that accompany them. Some feature several elements that refer to American culture. One example is the story of a 'down-and-out' baseball star who has to find his way again, and another features a cute blonde teenager who tries to attract the guy she likes (also a football player) whilst dealing with kids she has to babysit.

A typical stage has little more than 100 lines of text including captions, dialogue, and text embedded in graphic elements (e.g. newspaper titles and inscriptions, and most signs are translated).

By identifying some of the translation techniques adopted and the particular constraints faced during the localization process, the analysis examines how translators dealt with issues relating to adaptation, with special attention to the treatment of culture-specific items.

Before proceeding, it is appropriate to review some of the peculiarities of text in games and see how *EBA* functions in this respect.

In many games, users can pause and restart subtitles whilst playing (Mangiron and O'Hagan, 2006). Games share this trait with DVDs, but it is not the only form of control players have over the text. Often, a textual element remains on the screen until the player moves the game forward by inputting a command (Darolle, 2004). When single lines of text are supposed to appear gradually, players may be able to make them appear instantly or skip them altogether by pressing a button.

In older games, or more generally in games with basic graphics and audio, the absence of spoken dialogue makes it almost natural to set the pace of on-screen text according to a player's reading speed. However, the possibility of the player regulating the speed and persistence of dialogue and other types of text on screen exists even for many of the more technologically advanced games, where accelerating the speed of the display of subtitles often means abruptly interrupting spoken lines, as well as any facial animation and lip-synching.

This feature is inherent to the nature of the game medium. A combination of factors ranging from the genre of the game to choices made during its development determine the relevance of the many elements that make up a specific title, but each player attaches greater importance to those elements that are closest to his or her taste. This is by no means a characteristic unique to games, but is reinforced by their interactive nature. In role playing games (RPGs), a genre where plot and character development are key components, someone who was particularly interested in the story might choose to speak to all the characters encountered during the game whilst another player might mainly be interested in the action segments and cruise at great speed through the dialogue. Interactivity makes the video game an adaptive medium that adjusts to the needs and tastes of the user. In this sense, content creation does not end with the developers, but continues with the player: the player ultimately decides, to a degree dependent on the game, the final shape of the product.

In *EBA*, the player cannot influence the speed of in-game text display. The text is part of the comic panels that accompany the music and cannot be sped up or slowed down. Furthermore, the game cannot be paused to read the text. It can be paused, but this makes both screens go black until play resumes. This has obvious repercussions with regards to the translation, adding a time constraint to the spatial limitation determined by the flow of the songs and background stories. Note that this time constraint is quite different from that which exists in subtitling, where translators must take into account the three rhythms listed by Carroll (2004): "The visual rhythm of the film as defined by the cuts, the rhythm of the actors' speech and an audience reading rhythm." In *EBA*, the text is contained in comics that follow each other at a predetermined speed and have no connection with the songs.

The songs have no subtitles – interlingual or intralingual – and it is easy to understand why. Combined with the fact that there is a lot going on in the game, the small *DS* screens would make it nearly impossible to follow subtitles during play, and the song lyrics only loosely relate to the stories.

3.1.1 Borrowing from comics: the use of onomatopoeia

As the stories are narrated in the style of comics, the text features several onomatopoeic expressions. Given the major role of onomatopoeic and mimetic words in the Japanese language, the translation of onomatopoeia has received significant scholarly attention in Japanese translation studies (Flyxe, 2002; Inose, 2007; Lee and Shaw, 2006). Italian translation studies have researched onomatopoeia mainly in relation to its prominent role in comics.

Zanettin (1998) noted that Italian comics borrowed a number of onomatopoeic words (e.g., gulp, slam, sigh, and mumble) from English. Modifying visual signs containing English words that formed part of the image was often cost-prohibitive, so translators chose to retain them in order to eliminate the need to redraw the images from scratch. This use of existing English created a substantial repertoire that is used today even in comics that are not translations of English originals but are instead created by Italian authors for the Italian market. For this reason, several of these expressions now have an obvious meaning for someone who is at least slightly familiar with the medium, despite not being actual words in the target language. However, as Zanettin points out, sometimes terms that are part of graphic elements are left in the source language even if they do not belong to this established pool of expressions. This makes comics one of the most difficult literary forms to completely decipher, as the degree of understanding depends "on the level of knowledge that the reader of the translated comic has of the culture and language of the original text".

To analyze the use of onomatopoeia in *EBA*, we focus on a sample gathered from some of the stories; the numbering reflects the order they appear in the game.

	Original English	Italian translation
1	KA-CHNK! [car door closing]	KA-CHNK!
2	POOF! [object disappearing]	POOF!
3	GLEAM [treasure chest gleaming]	BRILL
4	ROLLIN' ROLLIN' [rolling boulder]	ROTOLA ROTOLA
5	RIP [the sound of a cloak being torn]	RIP
6	WHEEEE! [baby screaming, happy]	NGÀ!
7	YEE-HAAW! [kid running on a skateboard]	IH OOH!
8	WAAAH! [baby crying]	UAAAH!
9	WOOF! [exclamation of triumph]	URRÀ!
10	YUM YUM [child munching on donuts]	YUM YUM
11	SPLAT! [child hit by a flying hot-dog]	SPLAT!
12	PANT PANT [babysitter panting while running after children]	UFF
13	GAWAHAHAEHH [man daydreaming of his loved one]	SIIIIIIGH
14	SWICK [a quick brushstroke]	SWISH

Table 1: Examples of onomatopoeia in *Elite Beat Agents*.

Clearly, the translation adheres to no uniform rule. While some expressions are unchanged, others have, to differing degrees, been modified. Two of the examples (7 and 8) use the simplest form of adaptation – purely phonetic. 1, 2, 5, 10 and 11 are unchanged. In example 10 ('YUM YUM' – used to indicate someone enjoying food), the translator chose not to change the original text, even though the target language has an onomatopoeic expression ('GNAM GNAM') that uses a similar sound to convey exactly the same meaning. In examples 3, 4, 6, 9 and 12, equivalent Italian words or onomatopoeic expressions are substituted, even though number 12 is one of the above-mentioned visual signs that, originally borrowed from English, are now common in Italian comics. On the other hand, the fact that examples 13 and 14 have been translated using two English words is a sign of how deeply ingrained some of these expressions (e.g., 'SIGH') are in the target language.

This analysis shows that the treatment of onomatopoeia in the game varies. What is worth noting is that, while practical constraints (such as the impossibility of changing a single textual element without redrawing an entire panel) might dictate many translation choices in comics, this type of constraint is not as stringent in *Elite Beat Agents*: the game often superimposes text layers onto a background, in a typical example of the *multilayered graphic format* mentioned by Bernal-Merino (2007). In the case of onomatopoeia, this is clearly visible since many of these expressions appear as animations over background pictures that are themselves animated images. This method allows for less costly modifications, as it is possible to modify only those layers of an image which contain the text. As such, the decision not to translate some expressions indicates a desire to retain the feel of the original language.

3.1.2 Culture-specific items

Besides onomatopoeic expressions, *Elite Beat Agents* contains other types of text items that require a player to have a basic knowledge of English, or at least a certain familiarity with North American culture. Table 2 gives examples that show the approach to translating these items.

	Original English	Italian Version (together with its back-translation, when different from the original)
1	TOUCHDOWN!	ARRIVATI! [*ARRIVED!*]
2	Poncho, 33, policeman.	Poncho, 33, poliziotto.
3	BANZAI! BANZAI! [men jumping out of the way of a taxi]	BANZAI! BANZAI!
4	Star Receiver	giocatore di football [*football player*]
5	Touchdown!	Touchdown!
6	Kiss this receiver goodbye!	Di' addio al campione! [*Say goodbye to the champion!*]
7	RHOMBULAN DETENTION CENTER	CENTRO DI DETENZIONE ROMBULANO
8	Home Run Hitter	Giocatore di Baseball [*Baseball Player*]
9	Home Run Hero	Campione [*Champion*]
10	Romancing Meowzilla	Romantico Miaozilla [*Romantic Miaozilla*]

Table 2: Examples of cultural references in Elite Beat Agents.

The word 'touchdown' received different treatment in examples 1 and 5 because the first occurrence refers to a taxi arriving at its destination and the second to an actual touchdown performed by a football player, something that becomes clear when displaying the term alongside the images. All other references to American sports terminology (4, 6, 8 and 9) have been slightly adapted for the foreign audience, indicating a desire not to alienate Italian players by using obscure terms, even if this means losing some detail (e.g., a 'star receiver' becomes the generic 'football player').

This might suggest a tendency towards domestication as defined by Venuti (1995), given that complete domestication (for example substituting soccer, a sport that is much more familiar to an Italian audience, for American football or baseball) was not a viable option as it would have required intervening in the whole graphics compartment.

However, the rest of the sample shows a preference, noticeable throughout the game, for retaining the cultural references in the Italian translation. Fortunately, Italy and North America share a pool of cultural references, which originate from Italy importing a large number of cultural products from the US over the years. Consider, for instance, examples 2, 7 and 10.

In example 2, the name of the police officer (Poncho), is probably a nod to the character Ponch, a motorcycle police officer from the popular US TV series *CHiPs*. During the 1980s, the series was broadcast in Italy, and became quite well known. This meant that this name could be maintained in the Italian version. In examples 7 and 10, the Rhombulans (alien invaders whose name is a portmanteau of *rhombus* and the *Star Trek* alien race Romulans) and Meowzilla (a giant cat, obviously related to the Japanese monster *Godzilla*) became Rombulani and Miaozilla respectively. The translation of the cultural references in (7) and (10) posed little challenge due to the familiarity of both *Godzilla* and *Star Trek* in both the source and target cultures. Moreover, these puns were easy to render in Italian, as the translation could be almost identical in sound and meaning to the wordplay in the source language – the Italian *rombo* could be used for *rhombus* and *miao* (the Italian onomatopoeic word for a cat's meow) as a translation for the original *meow*.

Example 3 deserves separate consideration. In the English version, the word *banzai* is already a term borrowed from Japanese. In the Italian version the translator left it unchanged, judging (correctly, in the author's opinion) that its use would be received in a similar way by both Italian and North American audiences. As with the decision to maintain most cultural references, this choice was acceptable because of the closeness of the North American and Italian cultures and because globalization has produced a certain familiarity with Japanese culture that forms part of the common ground the two share. When this common ground does not exist, or is not as extensive, localization becomes considerably more challenging. This is the rationale behind the following section: what if translation, or even *transcreation*, is insufficient?

4. *Osu! Tatakae! Ouendan!*

In 2005, one year before the North American launch of *Elite Beat Agents*, Nintendo released a Nintendo DS rhythm game developed by iNiS solely for the Japanese market, called *Osu! Tatakae! Ouendan!*. The premise of the game was the same as for *EBA* – the player would, through music, bring aid to people in trouble in their everyday lives. These people, in their most desperate moments, would call the *Ouendan* (literally, 'cheer squad') for help. The game was quite successful in Japan, and Nintendo decided to produce a version for the Western market. However, because it featured many references specific to Japanese society, Nintendo could not directly export *Ouendan* as it was.

The first problem was that the main characters were a squad of deadly serious male cheerleaders. Although not uncommon in Japan, North Americans would not find these characters familiar. Therefore the developers needed a set of characters which were acceptable to a North American audience but which remained as faithful as possible to the feel of the originals. Thus, the male cheerleaders wearing traditional Japanese school outfits became a trio of black-suited, sunglass-wearing special agents[1].

Transforming the cheerleaders into special agents was only the first step in making the game accessible to a Western audience. The entire soundtrack was replaced, the new track list including songs by popular Western artists replacing the Japanese pop and rock hits[2]. The background stories required revision, since most were set in Japan and used many culture-specific references. One, for example, unfolded at a race during a summer *matsuri* festival, a celebration that only someone familiar with Japanese

1 When asked about the process of changing the lead characters, Keiichi Yano, vice president of development at iNiS, stated that he tried to think of figures from American culture that could be similar to male cheerleaders helping people in need: "The first couple of keywords that we had were Blues Brothers, Men in Black, and Ghostbusters [...]. I always knew they needed to be some kind of special task force, like the CIA". Online at: <www.1UP.com> (consulted 18.09.2006).

2 Like *Osu! Tatakae! Ouendan!*, which contains songs by L'Arc-en-Ciel, Kishidan and Orange Range, *EBA* features covers of well-known pop and rock songs. The artists included range from Madonna to Deep Purple, from Avril Lavigne and Destiny's Child to David Bowie, making a more eclectic selection of songs than in the Japanese original.

customs would know of. Another revolved around a *rōnin*[3], a term which indicates a student who has failed his college entrance examination and is spending the following year studying on his own to try to enter college later. Other characters in need of help included a pottery master who had lost his inspiration and a restaurant owner desperately trying to make *ramen* (noodle soup) that his customers would appreciate. When playing these levels, Western users would certainly not feel that the game had been created expressly for them. This would detract from their gaming experience. iNiS therefore opted to rewrite the stories. As the games were developed for the DS handheld console, the cost involved in drawing everything from scratch was not prohibitive, as it might have been for a game created for the more advanced PS3 or Xbox360 consoles.

The final game, *Elite Beat Agents*, changed all the elements which were considered too culture-specific. Some, like the *haikus* (a traditional form of Japanese poetry) that appeared at the end of every stage, were simply eliminated. Others, like most of the characters and situations, were adapted for the different culture.

However, as pointed out by Mangiron and O'Hagan (2004) and O'Hagan (2007), Western users sometimes "expect a distinctive original flavor to be retained". Acknowledging the role they played in the success of the original game, iNiS retained the characteristic humour and Japanese story design[4]. The new game did not eliminate references to Japanese culture, but instead filtered them through a Western perspective. For example, one stage designed specifically for *EBA* saw a young ninja trying to restore his family honor by stealing back the plans for a new car designed by his father's firm. To achieve this he had to face a series of situations similar to a parody of a Hollywood action movie. Although the ninja comes from Japanese tradition, Western players recognize the figure immediately, and in the game he operates in a familiar context.

Bernal-Merino (2006) identified the distinctive feature of video game translation in what he called "a shared authorship", stressing the advantages of a business model whereby the creative and localization departments work almost simultaneously and translation helps shape the game

3 In ancient Japan, *rōnin* was the name of a samurai without a lord; nowadays the term indicates a student without a school.

4 Keiichi Yano commented on the necessity of retaining the original design style: "If we did a Stan Lee kind of deal, [it] would completely lose its flavor" (from 1UP.com, 18/09/2006).

itself. While it is not always possible to implement this model, in this case the team behind the Japanese game (iNiS) also created the North American version. This was key to ensuring that the feel of the game was not lost in the transfer, since the developer could aim to reproduce the same quirky humor with different characters and situations, thus creating a similar experience for foreign players.

As this analysis pointed out, while trying to maintain an individual aesthetic style, the localization team was not afraid to modify many of the aspects of *Ouendan* that were not part of the core gameplay. The result is a game with new characters and stories, as well as a new soundtrack – undoubtedly one of the most relevant components in a rhythm game. Nevertheless, the Japanese game and the localized version feel strikingly similar, as the gameplay is virtually unchanged. This type of localization represents an application of the ludological theories that have emerged in game studies, which see the ludic aspect (represented by gameplay) as the distinctive feature of the video game medium.

That said, acknowledging the effectiveness of this type of localization does not mean invoking it as a universal standard. Although gameplay is always of the utmost importance, other elements contribute to the unique identity of a specific game. Such drastic changes would probably not be advisable when localizing a role playing game, as an RPG with new characters and a new plot could hardly be considered the same game. The best approach depends on several factors, not least of which are the genre and, ultimately, the specific game characteristics.

4.1 Remakes

Trying to categorize the move from *Ouendan* to *EBA* is not easy. *EBA* has been called the spiritual sequel to *Ouendan*, although the sequel label is justified by little more than its later release and a few small tweaks in the gameplay. It is perhaps more appropriate to see *EBA* as a remake.

Remakes are not a recent phenomenon. In a broad sense we could say that remakes, intended as the retelling of a story, have existed since the dawn of literature. A number of academic studies deal with intralingual remakes, such as adaptations from one medium to another (e.g., novels adapted to movies) or remakes of classic works of the same medium (e.g., remakes of old movies). Although an exhaustive review of the various types of remakes is beyond the scope of this paper, it is worth noting that remakes in the same medium are, not surprisingly, common in games. With technology advancing at an impressive rate, a mere aesthetic update of a game can make it look new[5]. In addition, the continuous development of gaming systems created the now common practice of remaking titles for new platforms with only minor updates. This makes older games available to an audience that did not have the chance to play them when they were first released[6].

From a translation studies point of view, a promising though not yet widely pursued course of research is the analysis of remakes as translations. In a multidimensional taxonomy of translations, Gottlieb (2005: 7) collocates remakes of foreign films amongst the *isosemiotic inspirational translations*[7]. "Instead of merely translating the verbal elements (as in dubbing and subtitling [...]), a remake transplants the entire film, setting and all, into the target culture. The resulting film may appear to be an original work, but as it is based on an existing storyline etc., it is indeed a translation" (ibid.: 10). While his taxonomy does not explicitly include video games, titles like *EBA* would clearly fit into this category, with one crucial difference: the storyline is not always the common link between the original game and a remake, given that a game does not necessarily have a storyline.

As an example of this kind of extreme localization, O'Hagan (2007) examined the case of *Tokimeki Memorial* (1994 onwards), a series of Japanese games remade for the US market. In that case, the concept of a *dating simulator* was about the only thing left of the original games in the localized versions; the US versions could be considered to be completely new games. The case of *Ouendan*/*EBA* is peculiar in that the latter retained the

5 *Metal Gear Solid* (1998) and *Resident Evil* (1996) are two famous Playstation games that appeared as remakes on the Nintendo GameGube a few years after their original release.

6 Remakes for portable systems (such as the Nintendo Game Boy and DS, the Sony PSP and, in recent years, mobile phones) are typical examples. As their small size implies obvious technological limitations, they lag a few years behind the big home-gaming systems, making them ideal platforms for transferring ('porting') successful old titles. Recent transfers include the iPhone versions of classic games such as *Street Fighter II* (1991), *Broken Sword* (1996) and *Doom* (1993).

7 Gottlieb (2005) uses the term *isosemiotic* for translations where source and target texts use the same channels, and he defines as *inspirational* (as opposed to *conventionalized*) those where the translators are given greater freedom.

concept, mechanics and general atmosphere of the former, but involved a complete overhaul of both the textual and audiovisual elements. It is interesting to note that one of the few changes in gameplay was an adjustment of the difficulty level, which was reduced for the North American and European releases. Although developers have different policies in this respect, as there is no universally accepted belief concerning whether Japanese or Western players prefer more challenging games, this practice is not uncommon. This is a further indication of the complexity of localization: translation of the text is but one of its many parts.

5. Conclusions

If we wished to use a straight line to represent the many degrees to which a work can be translated for a foreign market, remakes would be located near the end of this spectrum. Viewing a foreign remake as an extreme form of localization allows a comparison of the extent to which a game can be adapted for different audiences. Specifically, this paper presented examples of the changes made to localize a game for a country with a cultural background distant from that of the original (the North American localization of a Japanese game) and the much smaller changes needed when the source and target cultures share numerous common referents (the Italian localization of the North American version).

This analysis illustrated some of the peculiarities of the game medium. Studies of foreign remakes in film and literature focus mainly on the story as the common element being presented to a new audience. However, using a localization approach that can be defined as ludological, two games can provide the same experience for players even if the language, setting, soundtrack, cultural references, and even the story are new. They can provide the same experience *because* all these elements are new, as long as there are no changes in gameplay, which, for many genres, emerges as the central basis for localization.

References

Bernal-Merino, Miguel. 2006. 'On the Translation of Video Games' in *The Journal of Specialised Translation* 6: 22–36.

Bernal-Merino, Miguel. 2007. 'Challenges in the Translation of Video Games' in Mangiron, Carmen (ed.) *Localització de videojocs*. Special issue of *Tradumàtica* 5. Online at: <http://www.fti.uab.es/tradumatica/revista/index_05.htm> (consulted 30.01.2011).

Carroll, Mary. 2004. 'Focus on standards – Subtitling: Changing standards for new media' in *The Globalisation Insider* 13 (3.3). Online at: <http://www.translationdirectory.com/article422.htm> (consulted 28.01.2011)

Di Marco, Francesca. 2007. 'Cultural Localization: Orientation and Disorientation in Japanese Video Games' in Mangiron, Carmen (ed.) *Localització de videojocs*. Special issue of *Tradumàtica* 5. Online at: <http://www.fti.uab.es/tradumatica/revista/index_05.htm> (consulted 28.01.2011).

Darolle, Katrin. 2004. 'Challenges in videogames localization' in *LISA Newsletter Global Insider* 13 (3.5). Online at: <http://www.lisa.org/globalizationinsider/2004/>(consulted 27.01.2011).

Dietz, Frank. 2006. 'Issues in localizing computer games' in Dunne, Keiran J. (ed.) *Perspectives on Localization*. Amsterdam and Philadelphia: John Benjamins. 121–134.

Entertainment Software Association. 2010. 'Essential Facts about the Computer and Video Game Industry'. Online at: <http://www.theesa.com/facts/pdfs/ESA_Essential_Facts_2010.PDF> (consulted 28.01.2011).

Flyxe, Martin. 2002. 'Translation of Japanese onomatopoeia into Swedish with focus on lexicalization' in *Africa & Asia* 2: 54–73.

Frasca, Gonzalo. 1999. 'Ludology meets narratology: Similitude and differences between (video)games and narrative'. Online at <http://www.ludology.org/articles/ludology.htm> (consulted 28.01.2011).

Gottlieb, Henrik. 2005. 'Multidimensional Translation: Semantics turned Semiotics' in Gerzymisch-Arbogast, Heidrun and Sandra Nauert (eds) *MuTra 2005 – Challenges of Multidimensional Translation: Conference Proceedings*. Saarland University. 1–29.

Inose, Hiroko. 2007. 'Translating Japanese onomatopoeia and mimetic words', in Pym, Anthony and Alexander Perekrestenko (eds) *Translation Research Projects 1*. Tarragona: Intercultural Studies Group. 97–116.

Kohler, Chris. 2006. 'Cheer Squad: Why iNiS wants to make you happier' in *1UP.com* (18 September 2006). Online at <http://www.1up.com/features/cheer-squad?pager.offset=0> (consulted 28.01.2011).

Lee, Wood-hung and Shaw, Yomei. 2006. 'A Textual Comparison of Japanese and Chinese Editions of Manga: Translation as Cultural Hybridization' in *The International Journal of Comic Art* 8 (2): 34–55.

Localization Industry Standards Association. 2007. 'The Globalization Industry Primer'. Online at: <http://www.lisa.org/Globalization-indust.468.0.html> (consulted 28.01.2011).

Mandiberg, Stephen. 2009. 'Translation (is) Not Localization: Language in Gaming' in *Software / Platform Studies, Proceedings of the Digital Arts and Culture Conference, 2009*, University of California Irvine.

Mangiron, Carmen and O'Hagan, Minako. 2006. 'Game Localization: Unleashing Imagination with "Restricted" Translation' in *Journal Of Specialised Translation* 6: 10–21.

Mangiron, Carmen. 2010. 'The Importance of not Being Earnest: Translating Humour in Video Games' in Chiaro, Delia (ed.) *Translation, Humour and the Media*. London and New York: Continuum. 89–107.

O'Hagan, Minako. 2007. 'Video games as a new domain for translation research: From translating text to translating experience' in Mangiron, Carmen. (ed.) *Localització de videojocs*. Special issue of *Tradumàtica* 5. Online at: <http://www.fti.uab.es/tradumatica/revista/index_05.htm> (consulted 28.01.2011).

Miller, Daniel, Aaron Parecki and Sarah A. Douglas. 2007. 'Finger dance: a sound game for blind people'. In proceedings at *9th international ACM SIGACCESS conference on Computers and accessibility*. New York: ACM. 253–254.

Muñoz Sánchez, Pablo. 2009. 'Video Game Localization for Fans by Fans: The Case of Rom-hacking' in *The Journal of Internationalization and Localization* 1: 168–185.

Rollings, Andrew and Adams, Ernest. 2007. *Fundamentals of Game Design*. New Jersey: Pearson Prentice Hall.

Yuan, Bei and Folmer, Eelke. 2008. 'Blind hero: enabling guitar hero for the visually impaired' in Proceedings at *10th international ACM SIGACCESS conference on Computers and accessibility*. New York: ACM. 169–176.

Venuti, Lawrence. 1995. *The Translator's Invisibility: A History of Translation*. London and New York: Routledge.

Zanettin, Federico. 1998. 'Fumetti e traduzione multimediale', *in TRAlinea*, 1: 1–11. Online at <http://www.intralinea.it/intra/vol1/zanettin.htm> (consulted 29.01.2011).

Games

Bioware. 2007. *Mass Effect*. Microsoft Game Studios and Electronic Arts.

Capcom. 1991. Street Fighter II. Capcom.

Capcom. 1996. *Resident Evil*. Capcom.

Harmonix. 2007–2010. *Rock Band* series. MTV Games and Electronic Arts.

Harmonix (2005–2007), Neversoft (2007–2010) and Vicarious Visions (2007–2010). 2005–2010. *Guitar Hero* series. RedOctane (2005–2009) and Activision (2009–2010).

id Software. 1993. *Doom*. id Software.

iNiS. 2005. *Osu! Tatakae! Ouendan!*. Nintendo.

iNiS. 2006. *Elite Beat Agents*. Nintendo.

Konami. 1996. *Dance Dance Revolution*. Konami Computer Entertaiment.

Konami Computer Entertainment. 1994–2009. *Tokimeki Memorial* series. Konami.

Konami Computer Entertainment Japan. 1998. *Metal Gear Solid*. Konami.

Nana-On Sha. 1996. *PaRappa the Rapper*. Sony Computer Entertainment.

Naughty Dog. 2007. *Uncharted: Drake's Fortune*. Sony Computer Entertainment.

Revolution Software. 1996. *Broken Sword*. Virgin Interactive Entertainment.

Sonic Team. 1999. *Samba de Amigo*. Sega.

Square Enix (formerly Square). 1987–2012. *Final Fantasy* series. Square Enix.

Square Enix and Disney Interactive Studios. 2002. *Kingdom Hearts*. Square Enix (formerly Square and Square Electronic Arts) and Disney Interactive Studios (formerly Disney Interactive and Buena Vista Games).

ANNELIES VAN OERS

Translation Strategies and Video Game Translation: A Case Study of *Beyond Good and Evil*

In this case study, the translation shifts in the Dutch localised version of the video game Beyond Good and Evil *are described with the help of thirteen translation strategies. The main focus of the study is diegetic text i.e. text that is part of the fictional world of the video game.*

1. Introduction

Video game translation has not yet been so fortunate as to receive extensive academic interest. A few academics have ventured into this new world of translation that violates some of the most traditional translation dogmas, but all told, it largely remains a neglected discipline of translation. This is hardly surprising as the translation of video games is a fairly new branch of the ancient translation tree. Although video game translation is slowly but surely earning a spot in the academic limelight, translation strategies applied in this field of translation remain largely unexplored.

In this paper, I present a case study focusing on the Dutch localisation of the video game *Beyond Good and Evil* (Ubisoft, 2003). This title received a full text localisation, meaning that all on-screen text was translated and all audio dubbed into Dutch. The localisation of this game is analysed with the help of translation strategies formulated by Vinay and Darbelnet (1995), Grit (2004), and Mangiron and O'Hagan (2006). Since the translation strategies used in video games have largely been unexplored, the main objective of the study is to establish which translation strategy is predominant.

2. Research Methodology

The research method applied for the case study of *Beyond Good and Evil* consisted of three steps. First, all the strings belonging to the *non-diegetic category* of text were removed from the data. Although text is an important element of most video games, from a translation research perspective not all texts are equally interesting. Hence, a clear distinction between text types which are relevant and irrelevant to translation research is needed. One way to achieve this is by dividing the various texts into so-called *diegetic texts* and *non-diegetic texts. Diegesis* is a term originally coined by Genette and used for texts that tell a story, or in other words contain a narrative (Shen2005: 107). For video games, this means that texts which are part of the fictional game world and thus add something to the story are considered to be *diegetic text* and are therefore interesting research material. Other texts, such as user-interface texts within menus or games platform terminology messages were excluded from the research material. These texts are usually translated according to predefined guidelines and leave translators no room for creativity. The result is that these texts are not the most interesting research material. Most video game translators will be able to make an educated guess about the diegetic or non-diegetic nature of a unit of text, but this is not a reliable method. The visual context of the video game also needs to be taken into account. Playing the game or watching online walk-through movies both provide a more reliable way of identifying the nature of strings, and this was the method used for this study. It then became a matter of watching gameplay movies, looking up the texts that appeared in the game in the master file, and providing a diegetic or non-diegetic label for every string. After non-diegetic text was discarded, 28,812 Dutch words split across 3160 strings were left for the next step of the research.

The next step was to label every unit of text with a maximum of three translation strategies. The majority of text strings in *Beyond Good and Evil* are fairly short and it turned out that no more than three translation strategies were used per string. It must be noted that only unique translations qualified for the analysis of translation strategies; repetitive translations which would have distorted the research results were excluded. Take, for instance, the alien race the DomZ, a name that is borrowed in the Dutch translation. Since these characters are mentioned fairly frequently, the borrowing strategy would appear to have a very high frequency in the results even though it is the same translation repeated multiple times. This is not only the case with the borrowing of DomZ, but also for other frequently occurring names and objects. The first instance of a name or object is included in the count, with any subsequent occurrences excluded to avoid distorting the research results.

After all the strings had been labeled with appropriate translation strategies, the final step involved counting how often a translation strategy was applied. This was a matter of copying and pasting filtered columns between multiple Excel sheets. In the end, following the removal of duplicate strings, 2196 unique text units were labeled with 2723 translation strategies.

3. Translation Strategies

One of the best ways to analyse translation shifts between source text and target text is by means of translation strategies. For the analysis of the Dutch translation of *Beyond Good and Evil*, several translation strategies were selected from Vinay and Darbelnet (1995), Grit (2004) and Mangiron and O'Hagan (2006). For video game translation, the following translation strategies are most relevant according to the case study of *Beyond Good and Evil*. Translation strategies 1 to 4 belong to Vinay and Darbelnet, 5 to 7 to Grit and 8 to13 to Mangiron and O'Hagan:

	Translation Strategy	*Definition*
1.	Borrowing	Using a word from the source text untranslated in the target text
1a	Borrowing of names	Using a name of a character, place or object from the source text untranslated in the target text
2.	The Calque	Literally translating linguistic units of a word or expression
3.	Literal translation	Translating a source text word-for-word into a grammatically and idiomatically correct target text
4.	Adaptation	Changing cultural references in the source text into ones appropriate for the target text
5.	Description	Explaining rather than translating a foreign cultural element

6. Core translation	Expressing the general meaning of realia
7. Omission 8. Re-naming	Leaving out irrelevant realia Giving key terminology and characters new names
9. Contextualisation	Adding extra information to the target text which is either not present or only implicitly present in the source text
10. Re-creation of play on words	Creating dynamic equivalent translations of humorous elements
11. Deliberate use of regional expressions	Using expressions typical of the culture of the target language
12. Transcreation	Extremely free translation recreating a text rather than faithfully translating it
13. Compensation	Adding new elements to the target text to make up for loss through translation elsewhere

Table 1: An Overview of Relevant Translation Strategies for Video Game Translation.

3.1 Vinay and Darbelnet

Vinay and Darbelnet (1995) provided the first four translation strategies used to analyse *Beyond Good and Evil.*

Borrowing

The first strategy is *borrowing.* Vinay and Darbelnet (1995) consider a word from the source language to be borrowed when it has found its way untranslated into the target language. This is usually done in order "to overcome a lacuna" of a metalinguistic nature such as "an unknown concept" (Vinay and Darbelnet, 1995: 31). In video games, the borrowing strategy is twofold. It is used both in the traditional sense as Vinay and Darbelnet intended it i.e. to bridge a gap that cannot be filled with a word from the target language, but it is also used for words that could be translated with an existing word from the target language. A very common term used in video games is *power up.* For instance, to power up a weapon means to temporarily enhance its capabilities. Most frequently, power ups are objects or features which need to be found or earned. For instance, a gun that usually fires a single bullet may be able to fire two bullets for a while after improving it with a power up. Generally speaking, the English term *power up* is borrowed in the Dutch language as *power-up*[1]. There is no Dutch word that fully covers the meaning of *power up* as used in games and the term is thus left untranslated. A term for which it could be argued that there is a suitable translation in Dutch is *upgrade*. In video games, when something is upgraded, it is permanently improved. For example, armour that before an upgrade would deflect thirty percent of the damage dealt to the game character would after an upgrade shield its bearer from sixty percent of the damage. The Dutch word *opwaardering* could be considered to be an acceptable translation. Yet this is rarely used and *upgrade* is borrowed in the Dutch.[2] In some cases, terms have been used in English for so long (from the very first video games) that translating them would only confuse gamers. The best example of this would probably be the word *level*, which describes an episode, chapter or section of gameplay in a video game.

The strategy of borrowing is also used for the names of people or objects. In *Beyond Good and Evil*, the warmongering species the DomZ are called the same in the Dutch translation. In order to make a distinction between normal borrowing of ordinary words like *power up* and the use of untranslated or unadapted names of characters or places, the sub-strategy of *borrowing of names* was introduced in this case study.

The Calque

The *calque* is the second translation strategy of Vinay and Darbelnet (1995) relevant to video game translation. Just like the strategy of borrowing, a calque usually fills a gap created by an unknown concept in the target language. The difference is that with borrowing, a source text word is introduced into the target language unchanged, whilst a calque borrows the meaning of a word or expression by literally translating its components or adapting its spelling to the conventions of the target language. An example of an English to Dutch calque is a *memory card* ("geheugenkaart"). A Dutch to English calque would be *kanaal* (canal) or *dijk* (dyke). A video game often features a fictional world with objects and particular moves that are invented especially for that video game. Names need to be thought up for those new elements and these terms are often as new and creative as the objects, actions or locations they describe. These names and terms are

1 The dash is added to adhere to Dutch grammar.

2 In fact, *upgrade* is also used outside the field of video games and is included in *Het Groene Boekje*, the official list containing the correct spelling of Dutch words.

often combinations of existing words that are put together to form a new word with a particular meaning and purpose in the fictional video game world. An example of a calque is a so-called *neutralising cannon*, used in *Beyond Good and Evil* to destroy enemies. The Dutch equivalent of this term was devised by translating both words literally, resulting in the term *neutralisatiekanon*.

Literal translation

Literal translation is the third translation strategy in Vinay and Darbelnet's model (1995) that is also commonly found in the target text of a localised video game. A translation is considered to be a literal translation if the source text is transferred into a target text word by word, leaving room for grammatical changes that ensure the correctness of that text. The following translation taken from *Beyond Good and Evil* is a good example of a literal translation:

Source Text	*Target Text*
This peaceful mining planet in System 4 is now completely encircled by the DomZ Armada.	Deze vreedzame mijnplaneet in Stelsel 4 wordt nu volledig omcirkeld door de DomZ-vloot.

Table 2: An Example of Literal Translation.

Adaptation

The final translation strategy from Vinay and Darbelnet's model which is relevant to video game translation is *adaptation*. According to Vinay and Darbelnet (1995), adaptation is used to replace a certain foreign concept in the source language with one that is more appropriate or understandable in the target language, thus creating a form of "situational equivalence" (ibid: 39). It should be noted that adaptation is only used to adapt certain aspects of a foreign culture and should not be confused with Mangiron and O'Hagan's (2006) translation strategy of *transcreation*, which is the adaptation of a source text regardless of its cultural appropriateness. Transcreation could thus be viewed as an optional translation strategy and not an obligatory one as adaptation is considered to be here. In addition, transcreation is applied to enhance the target text when compared to the source text, even when a fairly literal translation would suffice. Adaptation is performed so that the target audience is not confronted with a foreign concept that they do not understand or that would not seem appropriate. Vinay and Darbelnet (1995) provide the example of a father kissing his daughter on the mouth. In many cultures (including in Dutch society) this would be considered unseemly and this situation could be replaced with three kisses on the cheeks in the Dutch translation. Other instances where adaptation would be fitting are when cultural elements within the source text would not be understood by the target audience. Even though nowadays many developers design their games with globalisation kept firmly in mind, and despite the fact that many video games are situated in fictional domains that sometimes have few ties to the real world, every so often specific elements indigenous to a particular culture find their way into a video game. In *Beyond Good and Evil*, the character Pey'j jokingly offers the protagonist Jade a *candy gram* when she opens the door to him. A candy gram is a box of sweets delivered to someone's home with a personal message. In the Dutch culture, a candy gram is not common, but something that *is* frequently sold on the doorstep are so-called *kinderpostzegels* (postage stamps sold by children for charity). Here, the typical American candy gram was adapted to a typically Dutch phenomenon.

3.2 Grit

Grit (2004) originally formulated his translation strategies for realia, which he considers to be expressions or terms that are specific to a particular culture. Although Grit only uses his translation techniques for realia, the shifts occurring to realia during translation are sometimes the same as those that can be seen during the translation of words, word groups or even entire sentences.

Description

The translation strategy of *description* is particularly useful when there is no direct translation available. A concept or an expression, whether described in the source language by just one word or several, is explained rather than directly translated in the target language (Grit, 2004). A nice example in *Beyond Good and Evil* of such a defining translation is the sentence uttered by Pey'j:

Source Text	*Target Text*
Mm... Gotta find the mechanism and sweet talk it.	Mm... Ik moet 't mechanisme vinden en hopen dat het een beetje wil meewerken.

Table 3: An Example of Description.

To "sweet talk" someone is to coax that person into doing something he or she does not want to do or does not feel like doing. There is no direct Dutch translation for *sweet talk* in the sense of convincing someone or something to cooperate. The Dutch translation literally means: *I must find the mechanism and hope that it will somewhat cooperate.* This is not an exact translation, but it more specifically describes what Pey'j wants to accomplish.

Core Translation

A second translation technique devised by Grit (2004) relevant to video game translation is *core translation.* This strategy is used when a translation only expresses the general overall meaning of the original text. A hypernym is often used to convey the core meaning of the source text (Grit 2004), thus generalising the meaning in the target text. In the following translation featured in *Beyond Good and Evil,* this strategy is seen at work:

Source Text	*Target Text*
My little marvels: the Jet-Boots. They run on home-made bio-carburant.	Mijn grote trots: de straallaarzen. Ze werken op zelfgemaakte biologische brandstof.

Table 4: An Example of Core Translation.

"Carburant" is a French word for fuel that is not commonly used in English. The translator opted for a general translation of this word instead of a fancy one by using the more common Dutch term of *brandstof* (fuel). The Dutch sentence now literally reads: My great pride: the Jet-boots. They work on home-made organic fuel.

Core translation is not only used at word level, since it also helps translators to avoid approaching the subject directly when the context is not perfectly clear, as in the following string from *Beyond Good and Evil*:

Source Text	*Target Text*
Not exactly a warm welcome. I'm sure the Rhinos have got somethin' to take care of him...	Niet wat je noemt een warm welkom. Ik vermoed dat de neushoorns iets hebben om dit op te lossen.

Table 5: An Example of Core Translation.

The context of the above source string does not provide any information as to why or how *he* should be taken care of. There is obviously some kind of problem that needs to be dealt with. The translator opted for a core translation by saying (in Dutch): "Not what you call a warm welcome. I suspect that the Rhinos have something to solve this." The pronoun *him* has been replaced with a more neutral *dit* ("this"), thus conveying the string's general gist i.e. a problematic situation that will be dealt with.

Omission

In rare cases, if translators are unable to find a suitable translation, they may decide that some of the information in the source text is irrelevant or feel that they need to omit certain elements because of restrictions on length. This translation strategy of leaving out source text elements is better known as *omission* (Grit, 2004). Although at first sight this might seem a translation loss, often either it provides an improvement on a badly written English string or the Dutch translation benefits from a more concise formulation. A good example of an omission is the following string, again taken from *Beyond Good and Evil*:

Source Text	*Target Text*
The gravity of Hillys is too strong. We need a stellar motor **to have enough power.**	De zwaartekracht van Hillys is te sterk. We hebben een sterrenmotor nodig.

Table 6: An Example of Omission.

In the above example, the emboldened words were omitted in Dutch, meaning it now literally reads: "The gravity of Hillys is too strong. We need a stellar motor." We can only guess as to why the translator opted to leave these words out. It is most likely that the translation was too long and needed to be shortened to comply with the length restrictions imposed. The first sentence states the problem (we can't leave Hillys) while the second sentence states the solution (we need a stellar motor), thus implying that such an engine will be powerful enough. The phrase *to have enough power* was thus considered redundant.

3.3 Mangiron and O'Hagan

Apart from the above general translation strategies that can be employed in any translation discipline, there are several approaches that are especially relevant for video game translation. Mangiron and O'Hagan (2006) have formulated a number of strategies which were implemented during the translation of the video game *Final Fantasy X*. These strategies are: the re-naming of key terminology and character names; contextualisation by addition; the re-creation of play on words; the use of regional expressions; transcreation; and compensation.

Re-naming of key terminology and character names

Mangiron and O'Hagan's first relevant strategy is the *"re-naming of key terminology and character names"* (Mangiron and O'Hagan, 2006: 17). According to the authors, "names used for weapons, items and abilities form essential key terminology in video games" (2006: 17). In addition to weapons, items and abilities, objects like armour, people, vehicles, locations, attacks, foods, power ups, upgrades etc. are also given rather inventive names by their creators. Since this key terminology plays an important role in many a fictional video game world, the translation of this terminology and labelling needs its own strategy in order to appropriately convey their important function in the target text. Re-naming is essentially giving new names to objects in a video game which are appropriate for the target language and culture. The re-naming strategy works for both key terminology describing things and for the names of characters. An example of such re-naming in *Beyond Good and Evil* would be a device called a POD used to fix hovercrafts. The abbreviation POD stands for Pocket Optimizer Drone. This nice abbreviation is also a factor to be considered whilst translating. In the Dutch version, it becomes a TOR, short for Techniek Oplap Robot (Technological Patch up Robot). Incidentally, *tor* also means *beetle* in Dutch. Thus, both in English and Dutch, the abbreviation is an existing word and one that indicates this drone/robot is something small.

Characters in video games are also often renamed. Mangiron and O'Hagan (2006) give the example of a Japanese character Jitan from *Final Fantasy* IX who was renamed Zidane for the American version and renamed yet again into Yitán for the Spanish version in order to avoid possible legal issues with the well-known French football player Zinedine Zidane. In *Beyond Good and Evil*, the name Kip was changed to Kim, as *kip* in Dutch also means *chicken*. This would have struck Dutch gamers as very odd, hence a more appropriate name was chosen.

Contextualisation by addition

A translation strategy also frequently seen in video games is *contextualisation by addition* (Mangiron & O'Hagan, 2006). This translation strategy involves adding extra words, sentences or information in the target text which are not present in the source text version, assuming of course that the context allows for such an addition and that the added text does not convey any false information. Often the implementation of this strategy is again left to the discretion of the video game translator, as there are no fixed rules that decide when this strategy is deemed to be appropriate. At times, this strategy is rather obviously implemented, as in a scene of *Beyond Good and Evil* in which the characters Pey'j and Jade pretend to operate on a patient when all they are actually doing is fixing a new fuse to a broken elevator:

Character	*Source Text*	*Target Text*
Pey'j	Ah ha! You're gonna be able to operate.	Ah ha! Dan kun je de transplantatie uitvoeren, zuster Jade.
Jade	Fuse in place, doctor.	Zekering is geplaatst, dokter.

Table 7: An Example of Contextualisation by Addition.

The translation of the Dutch version literally reads: "(Pey'j) Ah ha! Then you can proceed with the transplant, nurse Jade. (Jade) Fuse has been placed, doctor". In the Dutch version, the text *zuster Jade* (nurse Jade) was added to further enhance the medical metaphor.

At other times, the use of this strategy is more subtle, as in the following example:

Source Text	*Target Text*
Hey Jade, get a load of that! A DomZ meteor just made us a new skylight. Look up there!	Jade, moet je daar ,ns kijken! Zorgt een DomZ-meteoor toch nog voor een lichtpuntje, letterlijk, Kijk daarboven!

Table 8: An Example of Contextualisation by Addition.

The DomZ are a species which threaten the planet on which Jade lives. The planet is plagued by DomZ meteors which are wreaking havoc. The English string is fairly neutral. Pey'j says that a meteor has crashed into a building, thus making them a new window in the roof. In the Dutch translation, the entire skylight reference is omitted and instead the meteor is turned into a glimmer of hope. The literal translation of the Dutch text reads: "Jade, take a look at that! A DomZ meteor provides a bright spot, literally. Look up there!" The Dutch word *lichtpuntje* means both a speck of light or a bright spot but also carries the figurative meaning of a glimmer of hope or a light at the end of the tunnel. Thus, in the Dutch, an obviously negative object (the meteor) is the object of a positive play on words, thereby further emphasising the fact that the meteor is an unwanted and dangerous phenomenon in their world. Contextualisation by addition is sometimes not as subtle as in the above example concerning the skylight. Explicit information is sometimes added that can in no way be deducted from the source text alone. Contextualisation is the opposite of the strategies of omission and core translation because extra and more specific information is included instead of either being lost in a more generalised translation or being left out entirely. Contextualisation is often used either because the visual context asks for a more explicit translation or the gameplay needs further explanation. The following string is a the perfect example of where the target text is made more explicit than the source text:

Source Text	*Target Text*
Quick Fehn! Jump up!	Snel Fehn! Op m'n rug!

Table 9: An Example of Contextualisation by Addition.

The fairly general command of "jump up" has been translated into a more specific instruction: "Quick Fehn! Get on my back!" Thus, the source string is made more explicit by a contextualising addition.

The strategy of compensation also adds information to the target text which is not present in the original version. There is, however, a difference between the two strategies. The difference is that compensation adds a recurring element to a certain target string even though that element was not originally present in the source language version of that particular string to compensate for a loss in translation elsewhere. These elements can be a character's distinguishing way of expressing him or herself, a certain type of humour or any other type of recurring textual feature. Sometimes, jokes or a character's idiom is lost in translation and compensation is used to make up for that loss. The strategy of contextualisation adds extra bits of information to the target text on a more incidental basis and is not used to compensate for elements lost in translation.

Re-creation of plays on words

Re-creation of play on words (Mangiron and O'Hagan, 2006) might well be one of the most important translation strategies in video game translation. To put it simply: "humour is a key element in games" (Mangiron and O'Hagan, 2006: 18). The amusing effect is often achieved by either a play on words or just straight forward jokes, and serves to entertain gamers (Mangiron and O'Hagan, 2006). The first and foremost skopos of game localisation is to preserve the gameplay experience, and the entertaining effect of jokes is part of this. As a result, it is not possible to translate these puns and jokes literally, as this would mean that some or all of the humour would be lost. This is where the re-creation of plays on words is used. A joke or pun is not translated literally, but rather is replaced by an equally funny translation. To achieve this comic effect, the source text itself is considered less important and what counts is only what could be considered amusing in the target language. A good example of this translation strategy can be seen in the following string:

Source Text	*Target Text*
Well I'll be a monkey's uncle (nothing personal Jade), even the 'vators been screwed with...	Krijg nou tieten!! (Eh, sorry Jade)![3] Zelfs met de liften is gerotzooid...

Table 10: An Example of a Re-creation of a Play on Words.

3 It might seem strange that "Eh, sorry Jade!" is left in English, but in fact the word *sorry* is one of those English words that has found its way into the Dutch language untranslated and is now part of everyday Dutch idiom.

The character who utters this line is Pey'j, and he is talking to his niece, Jade. Those who have played *Beyond Good and Evil* might remember that Pey'j is a type of boar, with tusks and a belly to match. As such, Pey'j suggesting that he is related to a monkey is the first humorous element. Secondly, he is addressing his niece and thus insults her by basically calling her a monkey. The third meaning of this phrase is to express utter disbelief. The Dutch language does not have an equivalent expression and a literal translation would neither convey the humour of the original, nor the covert insult. The Dutch translation literally reads: "Get boobs now!! (Eh, sorry Jade!)..." In Dutch, this is quite a rude way to express surprise, but it fits perfectly well with Pey'j's personality and appearance. It is such a surprising remark (because of its rudeness) that it suddenly becomes funny, and even more so because Pey'j apologises to Jade afterwards. The source text is in no way retraceable in the target text, but the humour of this string is preserved.

The use of regional expressions

Another video game translation strategy is the use of *regional expressions* (Mangiron and O'Hagan, 2006). When a regional expression is used it means that a cultural reference to the target culture is deliberately added to the target text, which brings "the game closer to the audience" (Mangiron & O'Hagan, 2006: 19). This strategy is often adopted to give gamers the idea that a game was originally developed in their mother tongue in order that they are unaware of playing a translated game.

In *Beyond Good and Evil*, a famous Dutch commercial slogan is used as a regional expression:

Source Text	*Target Text*
WHEN MAMMA WONT GO, BETTER CALL MAMMAGOOOO!!	DAT IS NOU DE KRACHT VAN DE MAMMAGO

Table 11: An Example of a Regional Expression.

Dat is nou de kracht van de (that is the strength of) is a slogan that belonged to a large Dutch bank. Nearly everyone in the Netherlands would recognise this sentence. As an added bonus, the vocal actor who said this in the game was actually the same actor as voiced the commercial.

Sometimes these regional expressions are of a linguistic rather than a cultural nature. These replace a neutral translation that would also have fitted, but would have added less local flavour. This is seen in the following string, also from *Beyond Good and Evil*:

Source Text	*Target Text*
The regular army **doesn't have much to say in the matter** since the Alpha Sections arrived.	Het reguliere leger heeft **niets meer in de melk te brokkelen** sinds de komst van de Alpha Secties.

Table 12: An Example of a Regional Expression.

Here, a fairly neutral and literal translation such as *...heeft er niets meer over te zeggen* ("they have no say") or a more free translation as neutral as *...heeft geen invloed meer sinds* ("has no influence over") would have worked perfectly well, but the translator opted for the typically Dutch expression *...heeft niets meer in de melk te brokkelen* (literal translation: "have nothing to crumble into the milk"). This expression gives the text an authentically Dutch feel that would not have existed with a more neutral translation.

Although the strategy of using regional expressions has much in common with the strategy of adaptation (Vinay and Darbelnet, 1995), there is a difference between the two. Adaptation deals with specific cultural elements or concepts actually present in the source text that would not be understood by the target audience. The use of regional expressions goes further than adaptation. In a way, regional expressions are indeed used to replace cultural elements, but not because they would not be understood. Rather, the aim is to give a video game a "local touch and allow the players to enjoy it as if it were an original game" (Mangiron and O'Hagan, 2006: 19). What is more, these regional expressions are also used when no cultural element is present in the source text at all. Even if a fairly neutral translation would also be fitting, a regional expression would add that little bit of extra 'Dutchness' to a target text.

Transcreation

O'Hagan and Mangiron (2006) argue that because translators of video games are granted more freedom in their translation choices (albeit limited by several factors), the traditional term *translation* is insufficient. They therefore propose instead the use of the term *transcreation*. Generally speaking, video games translators do not faithfully translate the source text in a traditional 'translation science' sense, but rather rewrite or re-create the text into the target language. Sometimes this re-creation goes so far that there are no elements from the source text traceable in the target text. In this case study, this translation strategy was mainly used when no other translation strategy would adequately describe a translation shift.

Compensation

The final translation strategy is known as *compensation.* Mangiron and O'Hagan describe this translation strategy as: "introducing a new feature in the target text to compensate for a different one that could not be reproduced somewhere else in the text" (2006: 15). According to Mangiron (2007: 313), the generally accepted notion of translation loss is an "inevitable and intrinsic part of the translation process [...]". For the translation of video games, this loss has interesting consequences. Firstly, the many elements present in video games will let the gamer know that something has been lost in the translation process. Both the gameplay mechanics and the visuals are tell-tales of the gaps in the target text. Moreover, according to Di Marco (2007: 7) "a video game text cannot just lose meaning, fascination, humour and characterization" and therefore "a poor translation can have a negative effect on the gameplay experience" (Montón, 2007: 7). In order to avoid this loss, translators use the translation strategy of compensation. What exactly constitutes compensation can be best explained using a simple example. Should an English joke prove to be untranslatable into Dutch, the translator will create a target text that does not contain the joke, thus resulting in the translation loss of that particular joke. The view that no funny remark is preferable to displaying a pathetic sense of humour is the general opinion in the video game localisation industry. A few strings further along in the file, the English string contains no joke, yet the context and the situation allow for a Dutch joke to be made. In this case, the translator is free to add that joke and has thus compensated for the loss of the previous joke by adding a new one in a different location. Even though this example has been given for jokes, many other textual elements of video games are thus compensated for when the differences between the two languages cannot be bridged in order to recreate the exact intention of a particular string. For instance, in *Beyond Good and Evil* this translation strategy works particularly well with the translation of accents, as seen below in the lines of the Spanish character Secundo:

Source Text	*Target Text*
Una pearl d'Aramis. Bueno, Yade.	Een parel van Aramis. Bueno, Yade.
Units? You want units?	Geld? Jij wilt geld, si?
The director senora from the Science Center wants a **collection** of all the animals on the planet.	De senora directeur van het Centrum voor Wetenschap wil een **collectione** van alle dieren op de planeet.

Table 13: An Example of Compensation.

In the first string for instance, *una* was not translated into Dutch and could thus be marked as a translation loss. However, in the second string, the word *si* that was not present in the source string was added. The same goes for the word *collectione* in the third string. By using the translation strategy of compensation, the Spanish flavour of the character of Secundo is preserved even though the Spanish elements do not correspond exactly.

All authors discussed above are aware that many translation strategies are active concurrently. Vinay and Darbelnet (1995: 40) state that "some translations come under a whole complex of methods so that it is difficult to distinguish them." Grit (2004) also recognises that a combination of translation strategies occurs. In their case study, Mangiron and O'Hagan (2006) repeatedly explain that multiple translation strategies can be found in one source language and target language fragment they provide. It must be noted that it is very likely that other translation strategies can also be found in a translated video game. The set of strategies listed above has been selected as being especially relevant to video game localisation. These strategies do not focus on grammatical shifts or differences between language systems, but rather on shifts at a semantic level.

4. Main Findings

As table 14 shows, comprising a little over 65% of occurrences, literal translation is the predominant translation strategy for the diegetic text from *Beyond Good and Evil*. Transcreation is the second most frequent strategy with 10.76%. The translation strategies of the calque, adaptation, re-naming and compensation do not even reach the one percent level. The percentages for the remaining strategies: borrowing; borrowing of names;

description; core translation; omission; contextualisation; re-creation of plays on words; and deliberate use of regional expressions range in values from 1.03% to 5.14%.

	Translation Strategy	*Number of Occurrences*	*Percentage of Occurrences*
1.	Borrowing	119	4.37%
1a.	Borrowing of Names	140	5.14%
2.	The Calque	5	0.18%
3.	Literal Translation	1773	65.11%
4.	Adaptation	8	0.29%
5.	Description	46	1.69%
6.	Core Translation	53	1.95%
7.	Omission	82	3.01%
8.	Re-naming	15	0.55%
9.	Contextualisation	51	1.87%
10.	Re-creation of Plays on Words	28	1.03%
11.	Deliberate Use of Regional Expressions	99	3.64%
12.	Transcreation	293	10.76%
13.	Compensation	11	0.40%
	Total	*2723*	*100.00%*

Table 14: Translation Strategies found in Diegetic Text from *Beyond Good and Evil*.

It must be said that the results were quite surprising, when taking into account the fact that video game localisation is a translation discipline that thrives on a creative and free translation approach. A literal translation could be considered a fairly mundane translation strategy. There are, however, a few factors that might explain why the predominance of literal translation in *Beyond Good and Evil* is not altogether surprising. First of all, in terms of culture, the English/American and the Dutch share many characteristics that do not need explanations or adaptations when being translated. A second factor that should not be dismissed in this discussion of the predominance of literal translation is time. As Dietz points out (2007: 4) "games are often large projects with extremely tight deadlines". A translator/voiceover director[4] from the video game localisation field remarked

4 This translator/voiceover director works for the largest video game localisation company in the Netherlands and is a colleague of the author.

that "lack of time kills creativity and leads to more literal translations." Literal translations are the easiest, safest and quickest option, even though a freer, more time-consuming translation would lead to a better end result. Another possible reason as to why literal translation was so frequently found in *Beyond Good and Evil* lies in the text of the strings themselves. Not every string in this title contains a joke, pun, cultural concept typical of the source text territory or an object that needs re-naming. The majority of strings just contain ordinary text that needs no special translation strategy and is best suited to literal translation. If, for instance, only three puns were found in the entire game, it would seem logical that the translation strategy used for puns (i.e. the re-creation of plays on words) would also occur three times, thus resulting in a very low occurrence percentage for that particular translation strategy. If the total number of strings containing jokes makes up only one percent of all the strings in the video game, it therefore makes sense that the translation strategy for puns does not exceed that one percent. Taking this consideration into account, it is not surprising that some of the other translation strategies occurred so infrequently.

5. Conclusions

This case study focused on the Dutch localised version of *Beyond Good and Evil.* To provide more insight into a translated video game, the following translation strategies by Vinay and Darbelnet (1995), Gritt (2004) and Mangiron and O'Hagan (2006) were used: borrowing; borrowing of names; the calque; literal translation; adaptation; description; core translation; omission; re-naming; contextualisation; re-creation of plays on words; the deliberate use of regional expressions; transcreation; and compensation. These translation strategies were then used to describe the translation shifts between the *diegetic* source text and target text of *Beyond Good and Evil.* Comprising 65.11% of all applied strategies, literal translation was by far the most common of all the approaches. Taking into account the fact that video game localisation needs a more creative approach than, for instance, the translation of books or articles, the result that well over half of the diegetic text strings were translated literally was quite unexpected. This unexpected result could be explained by the fact that literal translation takes less time and is sometimes the easiest and safest option in a branch of translation famous for its tight deadlines. What is more, it could be that the majority of strings contain fairly standard source text that is not highly creative to begin with. In these cases, a literal translation suffices.

Although the results of this research were rather unexpected and quite surprising, the first tentative and exploratory steps into the world of translated video game analysis by means of a large set of translation strategies have been made. With the help of thirteen translation strategies taken from both the traditional translation science field and the latest developments in video game translation, the video game *Beyond Good and Evil* has revealed its well-guarded Dutch localisation secrets. This case study is fairly limited in the sense that it merely covers one game in one language pair with only just over a dozen translation strategies. It is the author's hope that many more will follow in her footsteps, further expanding the model and extracting more localisation secrets from other video games.

References

Dietz, Frank. 2007. 'How Difficult Can That Be?" – The Work of Computer and Video Game Localization' in *Revista Tradumàtica 5*. Online at: <http://www.fti.uab.es/tradumatica/revista/num5/articles/04/04central.htm> (consulted 24.06.2008).

Di Marco, Francesca. 2007. 'Cultural Localization: Orientation and Disorientation in Japanese Video Games' in (ed.) *Revista Tratumàtica 5*. Online at: <http://www.fti.uab.es/tradumatica/revista/num5/articles/06/06art.htm> (consulted 19.08.2008).

Grit, Diederik. 1997. 'De vertaling van realia' in Naaijkens, Ton and Cees Koster (eds). 2004 *Denken over vertalen*. Nijmgen: Vantilt. 279–286.

Mangiron, Carme. 2007. 'Video Games Localisation: Posing New Challenges to the Translator' in *Perspectives: Studies in Translatology*. 14 (4): 306–317.

Mangiron, Carme and O'Hagan, Minako.2006. 'Game Localisation: Unleashing Imagination with "Restricted" Translation' in *The Journal of Specialised Translation* 6: 10–21. Online at: <http://www.jostrans.org/issue06/art_ohagan.php> (consulted 30.09.2007).

Montón, D. 2007. 'It's a Funny Game' in *The Linguist 46.3*. Online at: <http://www.wordlabtranslations.com/download/its_a_funny_game_EN.pdf> (consulted 29.04.2013).

Shen, Dan. 2005. 'Diegesis' in Herman, David, Jahn, Manfred and Ryan, Marie-Laure. (eds). 2005. *Routledge Encyclopedia of Narrative Theory*. London: Routledge. 107–108.

Vinay, Jean-Paul and Darbelnet, Jean. 1995. *Comparative Stylistics of French and English: A Methodology for Translation*. Translated from French and edited by J. C. Sager and M. J. Hamel. Amsterdam / Philadelphia: John Benjamins.

GIANNA TARQUINI

Translating the Onscreen Text Blindfolded: Possibilities and Impossibilities

The current heyday of the mass consumption and global circulation of video games, along with their unquestionable cultural impact, has attracted the interest of a growing number of translation scholars. The specific nature of the medium and the sui generis industry-driven context determine a new area of specialisation, which to date remains largely unexplored in terms of its descriptive-analytical aspects. The investigation of translation behaviour, as advocated by descriptive translation studies (DTS), tends to be hindered by binding factors such as the non-standardisation of industrial practices, the textually evasive nature of the medium and the restricted access to primary materials. Drawing on the concept of "constrained translation" (Titford, 1982: 113), the present paper aims to highlight significant constraints and recurrent translation patterns in the approach to translating onscreen text (OST). These findings derive from a comparative pilot study conducted on a relatively large database of video game translation projects, in mainly the English-Italian and English-French language pairs, in the specific case of outsourced localisations. For methodological and procedural purposes, the translation analysis of the assets needs to be separated (into manual, dialogues, and onscreen text as described in this paper) and systematised according to relevant analytical categories, either borrowed from traditional DTS or motivated by specific localisation constraints. The final results, beyond highlighting regularities and possibilities, shed light on the difficulties and impossibilities determined by the current management of translation processes in game localisation.

1. Introduction

In the frame of audiovisual translation (hereafter referred to as AVT), the concept of *constrained translation* has gained ground as a fundamental assumption in indicating, firstly, that translation cannot be reduced to linguistic/cultural components and, secondly, that major translation problems "derive essentially from the constraints imposed on the translator by the medium itself" (Titford, 1982: 113). Although Titford coined this expression for subtitling, it has since been adapted to other audiovisual practices in subsequent research (Mayoral Asensio, Kelly and Gallardo, 1988) in building a semiotic taxonomy of the level of subordination to which AVT translation forms are subject (Diaz Cintas, 2004). Specifically, the concept of *constrained translation* refers to:

> situations in which the text to be translated is part of a more complex communicative event which attempts to convey a message by various means, such as pictures, drawings, music, etc. The translation of the linguistic part is crucially conditioned by the other elements and poses a set of problems that are specific enough to require special attention. (Bartrina and Espasa Borrás, 2005: 83)

Starting with this fundamental assumption, the present paper aims to place the concept of constrained translation in the specific field of game translation, empirically analysing the constraints and strategies imposed by the medium (and by the industrial treatment of game assets), in particular for the case of onscreen text (hereafter referred to as OST).

However, before examining the backbone of the study and its results, a review will be conducted of the fundamental factors affecting the approach to the translation of video games, such as asset types, game genres/platforms, and internationalisation/localisation management. This is a crucial distinction to make, since the management of internationalisation and localisation practices may determine different working conditions, and hence partially different translation constraints and strategies. The most symptomatic example in this respect is the in-house versus the outsourcing model.

After outlining these premises, the methodology of the comparative study which was carried out on 16 original/pivot English game scripts aligned to Italian and sometimes French translations after the proof-reading stage will be introduced. The focus will then shift to the structural/

textual features of the OST, which determine sui generis constraints on translation, related in particular to visual display, digitisation and file management, in line with the concept of constrained translation. These peculiar aspects will be discussed with the aid of significant examples extracted from the body of material which highlight recurrent problem-solving strategies, and recommendations also provided to game translators. In conclusion, whilst the "fun factor" has been widely acknowledged as being a guiding principle in game translation due to the communicative purposes of the medium (Mangiron and O'Hagan, 2006; Bernal Merino, 2011; Mangiron, 2010b), the present contribution brings to the fore the practical constraints within which creativity must be developed.

2. Fundamental Factors of Variation in the Approach to Game Translation

Translating video games requires the ability to cope with a variety of genres, text types and topics (Bernal Merino, 2007, 2008). From a broader perspective, and also for descriptive research purposes, it should be noted that translation constraints and the overall translation environment of video game scripts can vary according to a wide range of factors, including among others:

- game assets
- game genres
- gender/target age groups
- game platforms
- language pairs/locales
- internalization/localisation management

These factors can be considered relevant in the formulation of a general translation approach to video game texts, based on functional, extra-textual and operational considerations. For example, as far as game assets are concerned, translating a game manual will naturally require a different approach to the translation of subtitles because the textual features, types and functions are dissimilar. This list of factors was partly inspired by previous research into translation studies and game translation (Baker, 1995; Nord, 2000; Mangiron and O'Hagan, 2006; Bernal Merino, 2007, 2008; Chandler and O'Malley Deming, 2011) and by the translation analysis of the video game database referred to in this study (see subsequent section), with particular regard to consistent variation factors affecting translation behaviour. In the following paragraphs, each category will be briefly analysed, largely from a functional/skopos perspective.

Game assets mainly comprise artificial textual entities created by the industry stakeholders in order to manage digital multimedia content. Drawing on Bernal Merino's classification, nine different types of translation assets can be identified: manual, packaging, ReadMe files, the official web site, User Interface, online help, artwork, dialogues for dubbing, and dialogues for subtitling (2008: 34). ReadMe files, for example, exhibit clearly distinctive features in relation to subtitles or dubbing, and hence call for dissimilar translation strategies. For analytical purposes, the systematic investigation of translation regularities requires that textual instances, which are representative of a specific language variety, be separated (Baker, 1995: 229–30).

Like film genres, game genres[1] usually imply relevant variations in literary or specialised domains (action, adventure, combat, sports, music, racing, simulation and so forth). Some genres are highly marked in terms of their interaction/narrative techniques, locations and characters, such as the role playing games (RPG) based on the *Dungeons & Dragons* series[2]. This sub-genre involves a high level of customisation of playable characters and a fantasy cosmology of medieval locations, weapons and outfits. It must comply with strict glossaries and the fictional terminology of *Dungeons & Dragons* (including dozens of manuals of previous titles which

1 The concept of "genre" and its classifications have been at the heart of debate in folklore studies, literary studies, rhetoric, media studies and linguistics (for a complete discussion of the term see Swales, 1990: 33–52 and Mittel, 2004). According to Newman (2005: 12–13), the existing taxonomy of game genres is fuzzy and to a great extent industry-derived and text-based, while it would be possible to understand genres in terms of interpretive communities and audience expectations.

2 *Dungeons and Dragons* (D&D) is the original role playing game of medieval fantasy and adventure, in which players take on the role of an imaginary character defined by a set of statistics, supernatural powers and magical abilities. Originally published in 1974 as a table game, the series has inspired a variety of editions, novels, cartoons and video games over the years. Popular game titles include *Neverwinter Nights* (2002) and *The Temple of Elemental Evil* (2003) (Slavicsek and Baker, 2005).

are used as a reference), and is often developed for online game sessions. Other genres such as quiz games entail a peculiar game structure, and tend to be more realistic, humorous and rich in cultural references. Furthermore, many games are inspired by popular books, comics and films, and require a deep understanding and rendering of elements of popular culture. It can therefore be assumed that, depending on the game genre, "the text to be translated will require a proficient understanding of the jargon used and an accurate rendering of that particular terminology for the locale" (Bernal Merino, 2007: 3). That said, it should nevertheless be pointed out that the labelling of genres according to distinctive aesthetic, narrative and translation features is not always clear cut, partly because some genres (such as action and adventure games) tend to share very similar features and also because recent titles tend to bridge genres (e.g. quiz + sports games; action + simulation games) (Newman, 2005: 55–57).

The gender and target age group are relevant factors in the choice of a macro-approach to translation which will guide micro-strategies at the sentence level. These categories are often related to video game genres: pink games are targeted at girls, so that a specific lexicon is implied (girl toys, pets and clothing), and the register is also supposed to be adapted to a young audience. In the European context, the PEGI (Pan European Game Information) ratings are useful, but do not represent an exclusive indicator since the PEGI classifications have a restrictive rather than an inclusive function: a game rated 3 + is suitable for all ages (not only children), whilst a game rated 16 is overtly not suitable for children. This distinction is relevant insofar as the translation approach should be adapted to the target addressee:

> What is common practice in technical writing and advertising is not so unanimously accepted in translation. There are still quite a few scholars who consider "the source text" (whatever that may be) to be the yardstick by which they measure the quality of a translation, and although many of them agree to the above-mentioned principles of good writing in one moment, they criticize functionalist approaches to translation on asking "how do translators know who the target audience will be and what their expectations are". There is, indeed, good reason to ask this question, and we shall try to find an answer – or at least a method by which a translator may find an answer in a particular translation situation. (Nord 2000: 198)

Game platforms tend to affect the translation approach with regards to terminology compliance and asset features. Hardware manufacturers (such as Sony, Nintendo and Microsoft) require each localised version to comply with an agreed list of terms and error messages for their consoles and peripherals, which is, however, chiefly a matter of glossaries and pre-translated sentences (Chandler and O'Malley Deming, 2011). In addition, game platforms tend, to a certain extent, to determine the manual length, the textual/functional features of the onscreen text and the implementation of audiovisual files. Hand held consoles, for instance, do not incorporate extensive dialogues, but strings that serve the function of subtitles or comic bubbles, usually incorporated in the OST. Consider for example how the translation project of *Tomb Raider Underworld* (2009) for the Sony Playstation 2 and the Nintendo DS may differ. While the game genre and target group is substantially the same, there will be relevant variations in the terminology requirements (i.e. the Sony Playstation 2 vs. Nintendo DS glossaries), as well as in the way in which assets are organised/formed. In some cases then, the same OST script is used for two or three platforms, with minor add-ons containing a sequence of hardware instructions adapted to each; in other cases, the OST is instead partially re-written for each console.

Obviously, translation strategies in game localisation are strongly influenced by typical language and cultural issues that recur in specific language pairs. More specifically, it could be said that cross-continental adaptation may require a high degree of "transcreation" (Mangiron and O'Hagan, 2006) at the level of graphics and storyline, while the main European versions (FIGS i.e. French, Italian, German, and Spanish) tend to be more homogenous and are often managed simultaneously within the same translation spreadsheet. As a consequence, the problem of domesticating distant original content (such as the Japanese feel) is not always a major issue within European language pairs.

Finally, the upstream management of internationalisation and localisation has an enormous impact on translation in terms of working conditions (project time, cost and quality, translation tools, memories and glossaries, etc.), procedures (either in-house or outsourced), the translation environment (textual format, the presence of parts of the programming code, contextual information, etc.) and the general management of audiovisual assets with regard to time and space constraints. It is difficult to make generalisations on this point, because each company has developed *ad hoc* practices for localisation, files management and the preparation of translation materials. Here, *ad hoc* also means that there are no standard practices in the development, programming, and internationalisation of original games, so that localisation practices vary accordingly. This is actually a major issue in the video game industry, where steps should be taken towards the integration of localisation in the development process through early planning, cultural awareness and localisation-friendly code (Chandler and O'Malley Deming, 2011: 4–8; Edwards, 2011). One of the most evident project management decisions affecting translation is the choice between the in-house model (involving direct access to the source materials and improved communication with developers and publishers) and the outsourcing[3] model, which implies working for a language vendor (LV) in an often decontextualised translation environment. Be that as it may, the time-to-market dynamics of 'sim-ship' (simultaneous shipment) increasingly require blindfolded translation or translating the assets before they have been finalised in any corporate model.

3. The Study Methodology

Within the Translation Studies (TS) framework, theoretical assumptions, as well as applied areas of training and criticism, are backed up by the descriptive branch (DTS). The aim of this is to "explain and predict what translating and translations are and will be" (Holmes, 2002 [1972]: 180–192). In the relatively new field of game translation, descriptive research is hindered due to ontological reasons (textual formats and metadata), the non standard corporate management of processes and materials, binding copyright laws, and, more generally, by the lack of an established methodology for the linguistic analysis of new media. As a consequence, linguistic and inter-linguistic description has been limited to ground-breaking case studies (see Mangiron and O'Hagan, 2006; Machin and Van Leeuwen, 2007 and Ensslin, 2009).

The results reported in the present contribution were part of a larger study conducted by the author for her PhD dissertation, in which she set out to investigate the interdisciplinary framework of game translation

3 Outsourcing is referred to as "the procurement of products or services from sources external to the organization" (Papaioannou, 2002: 6). Corporate decisions regarding the procurement of localisations are core issues for the utility software and game localisation industry.

(Game Studies, GILT[4], AVT) and examine translation behaviour empirically, drawing on a body of material of 16 parallel video game scripts post proof-reading. These were collected by the author during her professional experience working for two companies which specialise in the provision of outsourced localisation services to the game publishing industry. The original database contained a number of FIGS translations aligned to English source/pivot versions, totalling several billion words, but the focus was particularly on the English-Italian and sometimes English-French combinations (the Spanish translations are be shown within this study, when available).

As far as the factors of variation in the translation analysis are concerned (section 1), the projects contained in the database covered a variety of game platforms (with the exception of mobile phones) and were representative of the action-adventure and role playing game genres, although they varied with respect to the target age/gender groups and internationalisation/localisation management. Due to relevant functional dissimilarities, the assets examined (manual, OST, dialogues) were analysed separately through the creation of three sub-databases, one for each asset type, showing relatively consistent textual features in terms of the purposes, functions and translation constraints. Conducted in the form of a product-oriented comparative analysis, the study aimed to spot recurrent translation problems and strategies in the practice of video game translation. When the proof-read versions were not finalised, but did feature comments and modifications completed by the reviewers as in a 'work in progress', it was possible to formulate process-oriented hypotheses which were also backed up by project glossaries, translators' comments, and other project files. The translations carried out by the author herself represented a small percentage of the total volume of translated words all of which has undergone one or two review stages, thus ensuring a large sampling of translation behaviour.

After inspecting the peculiar features of each asset (the manual, OST and audiovisual script) along with the recurrent translation problems/strategies within each of these assets, the findings were systematised according

4 Game translation needs to be framed within the broader global context of GILT (Globalisation, Internationalisation, Localisation and Translation) in order to adequately explain and predict translation phenomena. In this paper, we will opt for the term *localisation* to refer to technical factors or to the broader context that incorporates translation.

to two sets of analytical macro-categories: 1) medium-driven and localisation issues, and 2) traditional translation studies and audiovisual translation categories. The first set of issues was inspired by previous research into utility software and video game translation (Esselink, 2000; Microsoft Style Guide, 2007; Minazzi, 2007; Bernal Merino 2008, 2011; Chandler and O'Malley Deming, 2011) and partly driven by the database inspection. The second set of analytical categories was derived from traditional issues in TS and AVT such as culture-specific references, lexical changes and verbally expressed humour[5]. In particular, the category concerning cultural elements has been widely explored in TS literature (Nida, 1964; Venuti, 1995; and Katan, 2004 among others) and in AVT (Tomaszkiewicz, 2001; Ballester Casado, 2003; and Ramière, 2006 among others), and may involve rather heterogeneous elements: toponyms, anthroponyms, food, institutions, celebrities, allusions and so on. In general, the analysis of cultural elements within the projects under examination produced rather heterogeneous results, since foreignisation and domestication strategies were largely bound to the client policy and the specific cultural orientation of each project. In this paper, the focus is restricted to the first set of technical issues related in particular to the specificities of the medium and to the localisation workflow, which tend to highlight recurrent constraints and translation patterns, as will be shown in sections 3 and 4.

4. The OST: a Specific Form of Constrained Translation

The onscreen text, also referred to as the *user interface* (UI), *graphical user interface* (GUI) or *in-game text*, is particularly related to software functionality and comprises "areas of the game where the user can input or receive information. For example, the user can select a character from a list of choices, or get information about his character's health from a health bar indicator" (Chandler and O'Malley Deming, 2011: 346). It includes both non-diegetic text (such as statistics, options and menus) and diegetic elements (character/item descriptions, narratives) displayed in the video game

5 For an insightful overview of verbally expressed humour (VEH) in TS, AVT and game translation see Chiaro, 2010a and 2010b.

user interface, which are partly subject to audiovisual constraints such as subtitles. However, because a video game is a piece of software, such information must be programmed in an artificial language and incorporated within the game source code in order to be compiled and finally run by a hardware system. As part of this "rewriting" process, verbal language is fragmented into single "language strings" and scattered across lines of code so as to accommodate interactive textual clusters. For translation purposes, programmers then extract single UI linguistic elements from the game code and paste these into a spreadsheet next to the their corresponding ID (string identifier), as shown in the following tables (below). Accordingly, the onscreen text retains particular features related to programming, which will impose a further set of constraints on translators in addition to the audiovisual challenges. Unlike subtitles, the onscreen text is non-linear due to interactivity, and is fragmented into single language strings[6]. As a consequence, the OST can be considered as a sui generis textual entity that poses a set of linguistic, semiotic and technical constraints that are specific enough to require special attention, in line with the principle of constrained translation (Bartrina and Espasa Borrás, 2005: 83).

The following sections focus on such constraints and discuss their consequences by illustrating significant examples of recurrent translation patterns. This is complemented with suggestions and recommendations that emerge from the comparative study and from the project guidelines and comments contained in the body of materials. Drawing on this framework, the major constraints in the translation of the OST relate to:

- visual display → resulting in binding space constraints and formatting issues
- digitisation → the presence of parts of the game code, tags and variables
- file management → decontextualisation, non-linearity, heterogeneity and ambiguity (as discussed in the next section)

Undoubtedly, one of the major challenges faced by game translators is the issue of space. Because the OST sentences displayed on the screen are

6 Subtitles for video game dialogues are not usually contained in the onscreen text, but in cinematic and voiceover assets (Chandler and O'Malley Deming, 2011: 166–174, 176–184). As the main focus of the present contribution is the content of onscreen text and translation constraints, the typical issues related to subtitling (reading speed and time constraints, sentence chunking etc.) will not be covered (see Mangiron, 2010a for an introduction to game subtitling).

subject to severe visual constraints, translation strategies such as expansions, periphrases or notes cannot be used (Gambier, 2008). In practical terms, the main space restriction is an Excel cell in the translation environment, and the corresponding hypothetical margins of the containing dialog box, menu or screen in the end-user display. Quite often then, the translator is required to adhere to more binding constraints, such as the maximum number of lines per cell and the maximum number of characters per line (see table 1). In the translation from English into the major European locales, the cell margins represent more or less severe space constraint, since the translated sentences involve on average a 30% text expansion compared to the English source text (Esselink, 2000: 67; Chandler & O'Malley Deming, 2011: 5). As a general rule of thumb, the translations under examination here did not require much more space than the source text (allegedly as a precautionary measure) and had to adhere to more severe restrictions when specified. It follows that a major translation priority throughout the OST is condensation, resulting in ellipses, abbreviations and reductions.

nb lignes max	nb car max / ligne	ID	EN	IT
4	24	MJEX04	Memorise the colours in the drawing. Then touch the paint pots to dip your brush and touch the **area you want to colour.**	Memorizza i colori del disegno, tocca i barattoli per intingere il pennello e poi l'area da colorare.
4	24	MJEX05	Follow the outline of the drawings with the stylus. If you go over the line, you'll have to start **again.**	Traccia il contorno dei disegni con lo stilo. Se non ricalchi la linea dovrai ricominciare.
4	24	MJEX06	Blow into the microphone to make the paper aeroplane fly. Be sure not to hit the flies or the **ground!**	Soffia sul microfono per far volare l'aeroplano di carta. Schiva le mosche e il suolo!
4	24	MJEX07	Find the pieces of poster on the village green and put them back together by sliding the pieces **with your stylus.**	Cerca i pezzi dei manifesti nel parco e ricostruiscili facendoli scorrere con lo stilo.

Table 1: Space constraints in the OST.

In this script, drawn from a party game released in 2009 for the Nintendo DS, the maximum number of lines and characters per line is strictly specified (in columns 1 and 2 i.e. no more than 4 lines and 24 characters per line). The English text, which was not the source version of the project, already exceeded the space limit, as indicated by the characters highlighted in red. Considering the average text expansion of Italian translations, it could be expected that a further reduction in relation to the English text was an extremely problematic task. The Italian translator attempted to keep to the required sentence length by applying such strategies as the use of elliptical paraphrases *(Then touch the paint pots to dip your brush and touch the area you want to colour → tocca i barattoli per intingere il pennello e poi l'area da colorare; Be sure not to hit → schiva)* and anaphoric references *(put them back together by sliding the pieces with your stylus → ricostruiscili facendoli scorrere con lo stilo).* Space economy was not as strict in all the scripts examined, since size limits are specified by the client only when they are most binding in relation to the final screen. Even so, comparative evidence showed that translators generally avoided excessively expanding the source cell length, and that proof-readers often intervened to abbreviate longer sentences. Not all issues relating to space observed in the database were fixed, nor were they always solvable by translators and proof-readers as, presumably, was the case for the text in table 1. In such cases, the final check is passed to the testers, who can work with the final multimodal version and are responsible for (among other tasks) GUI "cosmetic bugs" such as text truncation and wrapping (Davis, 2011). It is interesting to note that the problem of space limitations, which is actually a major constraint for translators, could be easily avoided just by programming resizable interfaces or by taking into account the average text expansion for foreign locales. A few technical precautions in the design of the UI would make the work of game translators much simpler (Chandler and O'Malley Deming, 2011: 6).

The OST is not a traditional written text created in a human language like, for example, a piece of literature or any Word file. Instead, it is made up of language strings extracted from a formal language, containing tags, metadata and variables. Translators must handle such non-linguistic elements with extreme care and be able to decode them as well as understand their consequences in terms of their impact on language syntax and morphology (Bernal Merino, 2011). The comparative analysis of the database showed that translators and proof-readers took extreme care in relation to fonts, mark-up and formatting issues. Translators tended to carefully maintain the format of the source text, also avoiding using double spaces and blanks after the final character of each cell. Translators were also required to avoid introducing special characters such as smart quotes (“ ”) which may tamper with the game code and cause functional bugs, instead using plain quotes (" "). Table 2, extracted from the OST script of a PC role playing game released in 2007, is a typical example of tag treatment. In this case, the words included between the initial <cBOLD> and final </c> mark-up tags will be displayed in bold characters in the GUI, so that the sentence <cBOLD>DAMAGE</c> BASED ON THE CHOSEN STYLE will be **DAMAGE** BASED ON THE CHOSEN STYLE on the final screen. It should be noted here that the translator carefully maintained the same symbols, spacing and formatting. Indeed, this is not a secondary aspect to game translation: because the text strings will be reintegrated within a formal language, and as they will be finally displayed on a multimodal environment, the same symbols and tags of the source version must be accurately maintained. Tampering with tags or parts of the code contained within the strings may cause an incorrect final display or even system crashes. As stated by the Microsoft Style Guide (2007), linguistic adaptation should not cause technical hindrance at a functional level. We can assume that the digital format of the OST, in which the handling of tags and metadata is as important as translation adequacy, is in itself a major constraint to game translation. In fact, regardless of the adequacy of the translation, if metadata are altered, the translation may not be displayed to the foreign user at all, or may cause the system to crash. This is a further burden on game translators, who are responsible not only for the linguistic and cultural transfer, but also for functionality issues.

it svswd ssm.xml	34761.Description	<cBOLD>DAMAGE</c> BASED ON THE CHOSEN STYLE <cBOLD>CHANCE OF CRITICAL EFFECT SLEEP</c> +50% <cBOLD>CHANCE OF CRITICAL EFFECT INCINERATE</c> +50% <cBOLD>OPPONENT'S SILVER SENSITIVITY</c> -15% <cBOLD>Silver sword with two sun runes and a moon rune, used for fighting monsters. Ineffective against humans.</c> <cBOLD>DANNO</c> IN BASE ALLO STILE SCELTO <cBOLD>PROBABILITÀ DELL'EFFETTO CRITICO SONNO</c> +50% <cBOLD>PROBABILITÀ DELL'EFFETTO CRITICO INCENERIMENTO</c> +50% <cBOLD>VULNERABILITÀ DELL'AVVERSARIO ALLE ARMI D'ARGENTO</c> -15% <cBOLD>Spada d'argento con due rune lunari e una solare, usata per combattere i mostri. Non ha alcun effetto sugli umani.</c>

Table 2: The treatment of tags and metadata.

When the source code is not localisation-friendly, tags can even hinder the intelligibility of the source text, as shown in the example below:

Message-ID	English	Character	Char Limit	[illegible]	Feedback	Spanish	[illegible]	Feedback
SYSTEM_033	The last area you've been to is<L> Jungle Ruins "<C:2>The Fountain Guardian<C:1>" <L> You've cleared <C:2><F:0><C:1> stages	Johnny Style	3 × 36	Le dernier niveau où tu es allé est<L> Les ruines de la jungle "<C:2>Le gardien<L><C:2>de la font.<C:1>" Niveaux réussis <C:2><F:0><C:1>	#Code Defect# There are possible defects with the <C> codes. [illegible] we followed the rules indicated in the [illegible] file, and we can't see any possible defect here.	Último lugar visitado. Ruinas de la<L>Selva "<C:2>El Guardián de la Fuente<C:1>" <L> Has superado las escenas <C:2><F:0><C:1>	[illegible]	#Too Long Line# Count for each line (the line marked with "*" is length over) 32<L>*38<L>19 FIXED

Table 3: The consequences of poor internationalisation practices.

This extract is drawn from the OST script of a role playing game created for the Nintendo Wii and released in Japan in 2007 and in Europe in 2008. This can be considered to be a symptomatic example of poor internationalisation planning, particularly with regards to source code pre-arrangement for foreign locales. In fact, the OST script is constellated of tags and formatting, which are difficult to interpret and must certainly have posed a great number of space, semantic and morpho-syntactic challenges. The cells highlighted in rose and violet account for the amount of extra-work required to check and fix each string. We can hypothesise that the tag <L> refers to a line break, that <C:2> and <C:1> relates to formatting information, and that other tags stand for variables. These tags were presumably inserted because the Japanese version involved strict size constraints and the programming structure did not allow for resizing. The sentence in table 3 is actually a system message, which should reflect the speech style of the game character *Johnny Style* and use a maximum of 36 characters for each line and no more than three lines (the 'Char Limit' specified in the fourth column). The translators and proof-readers probably had to familiarise themselves with the most recurrent tags, and try to imagine their final output in order to count the characters contained in each line. For example, in the Italian translation, the proof-reader had to abbreviate the second line (The Fountain Guardian → Guardiano fontana) after the first submission, due to the excessive length of the second line (flagged with *38 in the feedback column). The strange sequence of tags <C:2><F:0><C:1> is probably a variable referring to the number of cleared stages, and can therefore be counted as one or two spaces in the whole line length. The French proof-read version reported an incorrect manipulation of tags, while the Spanish translation proceeded smoothly. This example clearly shows the huge difficulties involved in translating and reviewing scripts of games that have not been appropriately internationalised.

Addressing specifically the issue of variables, this is again a specific feature of natural language that undergoes the processing or "grammar" of formal languages. Variables are used in programming to customise data that can take different values according to the interaction of the player and to economise the programming methods. This issue has frequently been addressed in the game localisation literature (see Bernal, 2007 and 2011 and Wittner, 2011):

> ...most games allow players to choose their name, gender, nationality, etc. This means that translatable strings will need to incorporate 'variables' (similar to those used in mathematics or physics) for the game code to be able to take such data into account and present the right text correctly phrased. Variables can be used in many complex ways to enhance players' immersion by addressing them, their chosen profile and their performance directly. (Bernal, 2007: 6)

Essentially, variables can refer to numbers (as in table 3, which is the equivalent to saying "You have cleared *x* stages"), but can also refer to character names and items, posing grammatical problems of concordance in addition to contextualisation difficulties, as illustrated in the example below:

ORIGINAL	TRANSLATION
Keep on your toes, %PlayerName%!	Stai pronto, %PlayerName%!
<g>Acquired %s<w> (publisher note: %s = Item name)	<g>Oggetto acquisito: %s<w>

Table 4: Translation strategies in the treatment of variables.

These strings (comprising both the original and the translation) are respectively extracted from the script of a simulation game released in 2008 for the Nintendo DS and from a musical role playing game released for the Xbox 360 in the same year. In the first case, the translator opted for the translation *Stai pronto, %PlayerName%!* (Get ready, %PlayerName%!). However, the choice to make the gender of the player explicit through the use of the male adjective *pronto* (ready) may not be accurate, since it would not agree with a female player name. A common strategy in the translations from the database was to keep and, more often, to introduce gender neutral phrases in the presence of variables. In other cases, the project managers had created a separate column for female gender translation options. The second example is more problematic: depending on the play session, the variable %s may be replaced with a feminine or a masculine noun such as *spada* ("sword", feminine noun in Italian) or *scudo* ("shield", masculine noun), and the past participle *acquired* must still agree with the gender of the item. To avoid any possible incongruence, the Italian translator devised a linguistic strategy to elude gender concordance problems by making the hyperonym *oggetto* (item) explicit before the past participle, followed by a column and the variable: *<g>Oggetto acquisito: %s<w>*. Bernal Merino concludes:

> Unfortunately, there is no set way to indicate variables; this will depend on the SDKs (software development kits) used and on the lead programmer of each project, which means that different games designate variables in different ways. Translators have to be aware of which strings belong to the game code and which strings belong to localisable assets, since a mere character in the wrong place may cause the game to crash or even the computer to lock up. (2007: 5)

In fact, we have noted that the tags and variables used in the translation project relating to table 3 may appear rather unfamiliar even to the trained eye. However, it may be useful to list the most recurrent escape sequences[7] used in programming and contained in the analysed scripts:

Escape sequence	Name and Description
\b	Backspace – move cursor back one space on the current line
\t	Tab – move cursor to the next tab stop
\r	Carriage return – move cursor to the start of the current line
\n	New line or line feed
\"	Double quotation mark – to output a " character
\'	Single quotation mark – to output a , character
\% ; <%; %	Variable – sometimes followed by continuous characters and a closing character/tag[8]

Table 5: Escape sequences used in imperative programming and commonly found in OST scripts (adapted from McElligot, 2007[9]).

7 Escape sequences, usually introduced by a backslash character (\), are used to represent special characters within a language string.

8 Variables cannot be defined as escape sequences in strictly technical terms, but they can be inserted within the strings through escape sequences and special tags.

9 Private communication held during the course *Imperative Programming* within the Masters in Localisation Technology, University of Limerick.

These examples point to the fact that the OST strings, although keeping the apparent shape of natural language, actually spawn from formal languages and undergo electronic processing by localisation engineers and by the game system during gameplay. The translator should always keep in mind that the OST fragments belong to the source code, which has its own grammatical and syntactic rules, and should hence handle the mark-up and tags with extreme care. The translator is also supposed to be able to contextualise the language strings in their final output, and to be able to mentally preview their possible morpho-syntactic combinations. This is an extremely demanding task and is sometimes impossible, especially if variable descriptions or contextual information is not provided by developers or publishers. Unsurprisingly, a considerable number of queries and comments from translators contained within the body of material requested explanations concerning these issues.

5. The Binding Constraints of File Management: Decontextualisation, Non-linearity and Semantic Ambiguity

Due to its electronic nature, the OST cannot be considered a textual instance of natural language in theoretical or even practical terms. It includes textual fragments often isolated from their programming and linguistic co-text, as well as from their final visual display. By definition, the OST strings do not usually exhibit any syntactic or consistency ties between them: their syntax is actually multimodal, and lies in the output screen layout, changing at runtime and often causing the same string to be combined according to different patterns depending on interactivity. The main problem with this special status of OST strings is that their final meaning is multimodal and strictly bound to changing audiovisual configurations, while the environment of translation is often decontextualised. In addition, OST strings do not often recur in a linear sequence as far as their script co-text is concerned, which is a further source of semantic ambiguity in the interpretation of isolated sentences or words.

The OST involves heterogeneous information, listed on a spreadsheet largely according to the source code ordering or to core UI objects (system messages, tutorials, menu screens, etc.). In particular, the main text types (or string types, considering textual fragmentation) include: menus, controls and instructions, diegetic instructions[10], game narration, system messages, and descriptions of locations, items and characters. Within this non-sequential array of language strings, the translator, in common with someone working on an archaeological site, is supposed to reconstruct the fragments and elude possible ambiguities with regards to their meaning, textual functions and audiovisual referents. Below are illustrated significant examples of string types and the challenges they pose:

TXT FILE NAME	STRING ID	SOURCE LANGUAGE	ITALIAN	CONSTRAINTS & CONTEXT INFORMATION
astérix_french.txt	10	Club	Mazzata	Combo"
astérix_french.txt	11	Power-hammer / Compressor	Super martello / Compressore	Combo"
astérix_french.txt	12	Twister	Vortice	Combo"
astérix_french.txt	13	Twister Fusion	Vortice di fusione	Combo"
astérix_french.txt	14	Mole	Talpa	Combo"
astérix_french.txt	15	Time Distortion	Distorsione temporale	Combo"
astérix_french.txt	16	Instant Army	Armata immediata	Combo"
astérix_french.txt	17	Dogmatix Upgrade	Aggiornamento di Idefix	Combo"
astérix_french.txt	18	Atomic Spiral	Spirale atomica	Combo"
astérix_french.txt	19	Atomic Menhir	Menhir atomico	Combo"
astérix_french.txt	20	Stereo	Stereo	Audio"
astérix_french.txt	21	Mono	Mono	Audio"
astérix_french.txt	22	Configuration A	Configurazione A	Controller"
astérix_french.txt	23	Configuration B	Configurazione B	Controller"
astérix_french.txt	24	Configuration C	Configurazione C	Controller"
astérix_french.txt	25	Move	Movimento	Controller"
astérix_french.txt	26	Action	Azione	Controller"
astérix_french.txt	27	Jump	Salto	Controller"

Table 6: Menus, controls and instructions.

In this extract, drawn from an action game from the *Asterix and Obelix* series released in 2003, notably the language strings are grouped according to the main game controls and options (combo i.e. combination of actions, audio and controller). The final column on the right, providing contextual hints, is indicative of good internationalisation practice. As shown in this example, contextual information is extremely useful to translators for making sense of a cryptic list of items the audiovisual configuration and linguistic function of which is not clear within the script. In this extract, however, possible ambiguities were represented by the string *Club* and by the words *Move* and *Jump* (altered and highlighted in yellow by the proof-reader). As simple as these words may appear, their translation is subject to their final display. *Club*, for example, does not refer to a visually displayed item but to a "combo", thus designating an action *(mazzata)*

10 By diegetic instructions, we mean game instructions that are narrated through the game characters and plot, as opposed to technical user interface menus and commands.

instead of a weapon *(mazza)*. The words *Move* and *Jump* could either be verbs in the imperative form or nouns – in this case they stand for nouns because they refer to controller buttons, not to actions.

Another aspect of ambiguity relies on the heterogeneous non-linear nature of the strings, which can switch from one textual function to another, as illustrated in the following example:

q2112_sraymond.xml	310991.LocPhaseName	{0>Clue<}0{>Indizio<0}
q2112_sraymond.xml	310994.LocDescription	{0>What do I really care about someone's private life? I'm no monk myself. It's not enough evidence to jump at conclusions.<}0{>Cosa me ne importa della vita privata degli altri? Neanch'io sono un santo. Non è una prova sufficiente per trarre delle conclusioni affrettate.<0}
q2112_sraymond.xml	310997.LocShortDescription	{0>This fact should not lead to far-fetched conclusions.<}0{>Questo fatto non dovrebbe portare a delle conclusioni azzardate.<0}
stringdb.xml		{0>Step Over<}147{>Continua[MSOffice3]<0}
q2112_sraymond.xml	311012.LocPhase	{0>Night<}0{>Notte<0}
q2112_sraymond.xml	311015.LocPhaseName	{0>Fire<}97{>Fuoco<0}

Table 7: Diegetic instructions and game narration.

The example in table 7, extracted from a PC role playing game released in 2007, contains "game narration" in the form of comments about what the player has done and will have to do, as in a dialogic monologue. In contrast, the string *Step Over* is decontextualised from the co-text of game narration and is ambiguous in designating possible actions that the player is supposed to perform on the screen. The translator submitted a query to the language coordinator, who suggested that the string was a game command flagged by a different identifier (SDB Entry, presumably source database entry) in relation to the other co-text identifiers (LocDescription, LocPhase, etc., which allegedly stand for "LocationDescription" and "LocationPhase"). After checking the glossaries of the publisher, it was suggested that it should be translated using the verb *Continua*. The instances of multiple ambiguities in the projects under examination were most often solved by looking at paratextual information such as identifiers, and publisher's notes (when provided), or by examining translation memories. CAT tools such as SDL Trados are quite widespread throughout the game translation industry and can help to solve contextualisation issues as well as to improve consistency. When aids were insufficient or unavailable, the translators usually submitted queries to language coordinators or directly to the publisher/developer via QA (Quality Assurance) modules. In the mainstream outsourcing model, language coordinators and proof-readers are supposed to fix such issues, sometimes resorting to feedback from developers, but the difficulties caused by translation out of context in terms of processes and quality remain formidable. Visual CAT tools, labelled 'WYSIWYG' (What You See Is What You Get) since they allow for the final version to be seen during translation, are increasingly being implemented in the localisation of utility software but have still not been deployed in the gaming industry due to the technical complexity involved in the game code, intellectual property issues, and also because of a lack of interest. Ultimately, unsolved translation issues are passed to the testing phase, which involves a different set of priorities and does not represent the cure-all solution to blindfolded translation.

As already illustrated in section 1, video game hardware manufacturers have implemented a strict terminological policy in order to ensure that system messages and components are translated consistently across projects. The example in table 8 (below), extracted from an action game from the *Asterix and Obelix* series released in 2003, is a system message referring to the platform of the Nintendo GameCube, generated by the software-hardware interaction.

trc gamecube – anglais.txt	Insufficient space on Memory Card in Slot A. Asterix & Obelix XXL requires 1 file and %TRC_MC_NEED_FREE_SPACE% blocks to save.	Spazio insufficiente nella Memory Card (Scheda Memoria) nello Slot A. Per salvare Asterix & Obelix XXL sono necessari almeno 1 file e %TRC_MC_NEED_FREE_SPACE% blocchi vuoti.

Table 8: System messages.

This string was highlighted in yellow and presumably fixed by the proof-reader because it did not comply with the standard message list provided by Nintendo. Platform messages are strictly codified by hardware manufacturers, not only in their single terminological designations, but also as fixed sentences such as error messages or saving data information. This is an important aspect of the OST translation, because although platform messages are materially contained in the script produced by the software developers, they actually refer to a hardware system manufactured by a third-party stakeholder.

Finally, the following example, drawn from a title from the *Asterix and Obelix* series released in 2008, accounts for item descriptions. Again, these may pose semantic problems if the referent is not made visible in a glossary or else hinted at (how can you describe something that you cannot see?), and also pose gender/number concordance issues in the translation of isolated adjectives. For example, the adjective *Used* (in this instance highlighted in yellow by the author of this paper) can have four morpho-syntactic combinations in Italian (usato/usata/usati/usate), depending on the gender and number of the referent. In this case, the items (winged shoes) were listed in a logical order and specified in the second column to the left, so that the translator could infer the referent which was being hinted at.

Items.xml	ITEM1/NAME	Winged shoes	Scarpe alate
Items.xml	ITEM1/SHORT_DESCRIPTION	Worn	Logorate
Items.xml	ITEM1/DESCRIPTION	Shoes built for speed not comfort, but don't expect to beat any records.	Sono costruite per la velocità e non per la comodità, ma non aspettarti di battere qualche record.
Items.xml	ITEM2/NAME	Winged shoes	Scarpe alate
Items.xml	ITEM2/SHORT_DESCRIPTION	Used	Usate
Items.xml	ITEM2/DESCRIPTION	This elegant pair will improve your speed and give you that flying feeling.	Questo elegante paio di scarpe migliorerà la tua velocità e ti darà l'impressione di volare.

Table 9: Descriptions of locations, items and characters.

In audiovisual translation, the central issue of the interplay between the linguistic code and its multisemiotic context has been widely studied (see Delabastita, 1989; Niemeier, 1991; Herbst, 1995; Bollettieri Bosinelli et al. 2000; and Chaume Varela, 2004), and significant steps have been taken towards the provision of complete source materials for translators/adaptors in order to tackle specific multimedia and pragmatic challenges. Also in this respect, decontextualisation remains a huge industry-driven problem within game localisation. Professionals and scholars are increasingly sensitising the community of developers to providing the full source game or at least contextual information to game translators so that they can properly understand the original content before translating.

6. Conclusions

The translation of video games is without doubt a form of specialised translation, which requires the ability to cope with a wide range of topics, text types and technical as well as audiovisual challenges. Beyond the scope of creativity (and the fun factor), in this contribution we have explored video game translation as a specific form of constrained translation, conditioned in particular by software functionality, the treatment of electronic data and the audiovisual interactive context. Furthermore, as highlighted in section 1, the translation approach to video games is bound to other different factors, meaning that the results of this study can be considered as early general conclusions in a specific language pair (English-Italian) with regards to the onscreen text in the industrial outsourcing model. Further research is required into other game assets, genres, language pairs, projects and into different translation management contexts.

Despite the intrinsic limitations in analysing translation constraints and patterns in the specific domain of game translation (due largely to the restricted access to video game scripts), descriptive research is crucial both for supporting theoretical assumptions concerning game translation theory and for applied purposes in order to prepare students for real-life professional practice. In this sense, primary materials are a fundamental ingredient of a translator's training, and practical examples can help trainees discuss the possibilities and impossibilities of the current translation management model from a critical perspective. Finally, descriptive data can contribute to raising stakeholder awareness concerning what blindfolded translation practically entails, and also to sensitising scholars and professionals working in different domains to the challenges of translating video games.

References

Baker, Mona 1995. 'Corpora in Translation Studies: An overview and some suggestions for future research' in *Target* (7) 2: 223–243.

Bernal Merino, Miguel Á. 2007. 'Challenges in the Translation of Video Games' in *Tradumatica* 5.1–7. Online at: <www.fti.uab.es/tradumatica/revista/num5/articles/02/02art.htm> (consulted 15.06.2013).

Bernal Merino, Miguel Á. 2008. 'What's in a Game?' in *Localisation Focus,* 6, 1. 29–38.

Bernal Merino, Miguel Á. 2011. 'Working with Translators' in Chandler, Heather Maxwell and Stephanie O'Malley Deming, S. (eds) in *The Game Localization Handbook*, (Second Edition) Massachusetts: Jones & Bartlett Learning. 107–118.

Ballester Casado, Ana. 2003. 'La traducción de referencias culturales en el doblaje: El caso de American Beauty (Sam Mendes, 1999)', in *Sendebar*, 14: 77–96.

Bartrina, Francesca and Espasa Borrás, Eva. 2005. 'Audiovisual translation' in Tennent, Martha (ed.) *Training for the New Millennium*, Amsterdam/Philadelphia: John Benjamins, 83–100.

Bollettieri Bosinelli, R. M., *et al.* 2000. *La traduzione multimediale: quale traduzione* per *quale testo?* Bologna: CLUEB.

Chandler, Heather Maxwell and O'Malley Deming, Stephanie. 2011. *The Game Localization Handbook* (Second Edition) Massachusetts: Jones & Bartlett Learning.

Chaume Varela, Frederic. 2004. *Cine y traducción.* Madrid: Cátedra.

Chiaro, Delia 2010a. *Translation, Humour and Literature.* (Volume I). London: Continuum Books.

Chiaro, Delia 2010b. *Translation, Humour and the Media* (Volume II). London: Continuum Books.

Davis, Alan. 2011. 'Games Localisation Testing'. Paper presented at *Localisation Research Centre Summer School in Computer & video game localisation.* (University of Limerick, 30 May–2 June year??). Online at: <http://www.localisation.ie/resources/courses/summerschools/2011/AlanDavis_Test.pdf> (consulted 12.12.2011).

Delabastita, Dirk. 1989. 'Translation and Mass Communication: Film and TV Translation as Evidence of Cultural Dynamics' in *Babel* (35) 4: 193–218.

Diaz Cintas, Jorge. 2004. 'Subtitling: the long journey to academic acknowledgement' in *The Journal of Specialised Translation,* 1. Online at: <http://www.jostrans.org/issue01/art_diaz_cintas.php> (consulted 04.07.2013).

Edwards, Kate. 2011. 'Levels of game culturalization' in *Multilingual Computing.* September 2011: 18–19.

Ensslin, Astrid. 2010. 'Black and White: Language ideologies in computer game discourse' in Johnson, Sally and Tommaso M. Milani (eds). *Language Ideologies and Media Discourse: Texts, Practices, Policies.*London: Continuum. 205–222.

Esselink, Bert. 2000. *A Practical Guide to Localization.* Amsterdam/Philadelphia: John Benjamins.

Gambier, Yves. 2008. 'Recent developments and challenges in audiovisual translation research' in Chiaro, Delia, Christiane Heiss and Chiara Bucaria (eds). *Between Text and*

Image. Updating research in screen translation. Amsterdam/Philadelphia: John Benjamins. 11–33.

Herbst, Thomas. 1995. 'People Do Not Talk in Sentences. Dubbing and the Idiom Principle' in *Translatio Nouvelles of FIT-FIT Newsletter*, 15, 3–4. 257–271.

Holmes, James S. 2002. 'The Name and Nature of Translation Studies' in Venuti, Lawrence (ed.) *The Translation Studies Reader*. London Routledge. 180–192.

Katan David. 2004. *Translating Cultures: An Introduction for Translators, Interpreters and Mediators*. Manchester: St. Jerome.

Machin, David and van Leeuwen, Theo. 2007. *Global Media Discourse: a Critical Introduction*. London: Routledge.

Mangiron, Carme and O'Hagan, Minako. 2006. 'Game Localisation: Unleashing Imagination with Restricted Translation' in *The Journal Of Specialised Translation*, 6. Online at: <www.jostrans.org/issue06/art_ohagan.php> (consulted 25.06.2013).

Mangiron, Carme. 2010a. 'Subtitling in game localisation: First steps towards developing a standard'. Paper presented at *International Conference Languages and the Media, (Berlin, 6–8 October 2010) and contained in the book of abstracts.*

Mangiron, Carme. 2010b. 'The Importance of Not Being Earnest: Translating Humour in Video Games' in Chiaro, Delia (ed.) *Translation Humour and the Media*, Vol. II. London: Continuum Books. 89–107.

Mayoral Asensio, Roberto, Dorothy Kelly and Natividad Gallardo. 1988. 'Concept of constrained translation. Non-linguistic perspectives of translation' in *Meta*, 33, 3.356–367.

Microsoft Language Excellence Team. 2007. *Microsoft Style Guide (English-Italian)*. Online at: <http://www.microsoft.com/language/en/us/download.mspx> (consulted 10.06. 2013).

Minazzi, Fabio. 2007. 'Lezione glossari e terminologia' in *Corso in internazionalizzazione e localizzazione del software*, Webcen. Online at: <http://webcen.dsi.unimi.it/wcinfo/index_corsi.php?corso=77501&anno_acc> (consulted 04.07.2013).

Mittel, Jason. 2004. *Genre and Television: From Cop Shows to Cartoons in American Culture*. London/New York: Routledge.

Newman, James. 2005. *Videogames*. London: Routledge.

Nida, Eugene. A. 1964. *Towards a Science of Translation*. Leiden: E.J. Brill.

Niemeier, Susanne. 1991. 'Intercultural dimensions of pragmatics in film synchronisation' in Blommuert, Jan and Jef Verschueren (eds). *The Pragmatics of International and Intercultural Communication*, Amsterdam/Philadelphia: John Benjamins. 145–162.

Nord, Christiane. 2000. 'What do We Know About the Target-Text Receiver?' in Allison, Beeby, Doris Ensinger and Marisa Presas (eds). *Investigating Translation*. Amsterdam/Philadelphia: John Benjamins. 195–212.

Papaioannou, John. 2002. *The Localisation Outsourcing decision*. MBA Thesis. University??. Online at: <http://lexworks.com/cms/wp-content/uploads/2012/07/MBA_Thesis_JohnPapaioannou.pdf> (consulted 13.06.2013).

Ramière, Nathalie. 2006. 'Reaching a Foreign Audience: Cultural Transfers in Audiovisual Translation' in *The Journal of Specialised Translation*, 6, <http://www.jostrans.org/issue06/art_ramiere.php> (consulted 13.11.2011).

Slavicsek, Bill and Baker, Rich. 2005. *Dungeons and Dragons for Dummies.* Indianapolis: Wiley Publishing, Inc.

Swales John M. 1990. *Genre Analysis. English in Academic and Research Settings.* Cambridge: Cambridge University Press.

Titford, Christopher. 1982. 'Subtitling – Constrained Translation' in *Lebenden Sprachen,* 27, 3: 113–116.

Tomaszkiewicz, Teresa. 2001. 'Transfert des références culturelles dans les sous-titres filmiques' in Gambier, Yves and Henrik Gottlieb (eds). *(Multi)Media Translation: Concepts, Practices, and Research.*, Amsterdam/Philadelphia: John Benjamins. 237–247.

Venuti, Lawrence. 1995. *The Translator's Invisibility* London: Routledge.

Wittner, Judith. 2011. 'Improving translation of variables in interactive games' in *Multilingual Computing,* September 2011. 26–29.

Rafael Müller Galhardi

Video Game and Fan Translation: A Case Study of *Chrono Trigger*

Nowadays, video games are not restricted to a specific context or demographic group but instead involve an entire global entertainment market, in a similar manner to the film industry. Certain video game genres –largely RPGs (Role Playing Games) and strategy games – also rely strongly on textual components. Together, these two factors have generated significant demand for linguistic transference and differing degrees of adaptation to many specific markets – a process also known as localization. *However, for many locales, the localization industry has not yet reached its full potential, leaving the door open for amateur attempts at translating games. This article provides a descriptive analysis of the Brazilian fan translations of Square Enix's* Chrono Trigger, *which were based on the first official English translation (1995) carried out for Nintendo by Ted Woolsey, as well a comparison of these with another translation developed by Anglophone fans. The attempts at translation by these fans were possibly the motivation for a second official translation for the porting of the game to Nintendo DS (2008), which in turn generated yet another Brazilian fan translation, totaling a final group of six target texts. In order to conduct this research, Mangiron and O'Hagan's transcreation model (2006) was used in order to analyze topics such as: dialogue additions and omissions; the re-creation of play on words; the renaming of characters and terminology; censored items; the deliberate use of regional expressions; and even the modification of a character's speech style, in addition to any other challenges that* Chrono Trigger *presented to the fan translators.*

1. Introduction

Video games have become a global phenomenon, spreading in multiple directions and permeating many different languages. They have been generating significant demand for linguistic transference and a need for differing degrees of adaptation to many markets since the 1990s (Bernal, 2006: 23–24), in order that each player can experience and enjoy a game to its full potential. However, the industry cannot translate or localize every single game sold around the world, a fact that allows room for a growing practice in many different countries: the practice of fan translation.

Fan translation is not an isolated trend; it encompasses a great range of media production – from anime and literature to television series, movies and video games – and fan translations are distributed by a huge number of online communities. Muñoz Sánchez, a professional translator and experienced romhacker, explains that fan translation began with videogames: "Some fans started to study the structure of the ROM data of their favourite games, developing tools everybody could use to do fan translations. These are the origins of the so-called *'romhacking*" (2009: 169).

Today, there are many different groups appearing all over the internet seeking to gather fans in order to realize large-scale projects, such as the translation of games which were never released in the West, for example *Dragon Quest IV* (1990) or *Star Ocean (*1996), both translated by the group *Dejap Translations.* The translation process sometimes implies reengineering the game data so that Latin characters can be inserted. The Spanish fan translation of *Tales of the Abyss* (2005) is an example of the collective effort of fans, due to its great volume of textual content. Many translations, however, begin with the interest of just a single person. This occurred both with the retranslation of *Chrono Trigger*'s (1995) entire Japanese script by fan translator KWhazit (2007) – (later organized into Excel files by Zeality from the group *Chronocompendium*) and with the respective Brazilian fan translation carried out by Lynx (2004).

Fan translations are a reflection of what Jenkins calls *convergence culture*, meaning "a move from medium-specific content towards content that flows across multiple media channels" (Jenkins, 2006: 243). Fan translations fall into a legal grey area, since litigation against fan production has not yet occurred beyond the sending of so-called *Cease and Desist* letters. Companies' stances regarding fan activities are divided into two categories by Jenkins (ibid.: 134): the *prohibitionists*, for whom any manifestation of this grassroots creativity is potentially dangerous, and the *collaborators*, being those companies who harness fan enthusiasm and actively work with fan communities to promote their products.

Sometimes fan translations are the only option in terms of providing accessibility to a global audience, as evidenced by *Seiken Densetsu 3* (1995) and *Final Fantasy V* (1992). According to Muscar (2007: 252), "without the exposure of games like *Final Fantasy V* brought about by fan translations, Square Enix may have never released an official version because it may not have thought there was enough demand." However, it is still unclear whether fan translations utilize similar or very different procedures and translation strategies to their official counterparts, due to a lack of descriptive studies. Most of the time, fan translations only appear when there is no official localization available or in production, but on a few occasions, they are reinterpretations of an original game script, such as the work produced by *Chronocompendium* (2007).

Video game localization remains a very specific field; it differs from traditional software localization because it does not deal with what Thayer and Kolko call *productivity software,* that is "any software program that is primarily designed to facilitate work, whereas digital games primarily facilitate play" (2004: 3). Therefore, one needs to be aware of how videogames mix technical data with narrative and content issues relating to the core script and character dialogues in order that the end-product becomes not only functional but also enjoyable in the target language. In the words of Mangiron:

> It is therefore crucial that localisers are familiar with the localization process and the different kind of assets that form a game and possess the skills required to translate them, as games are made up of a diversity of texts, such as technical messages, witty dialogues, songs and literary passages, which have to be translated differently. (Mangiron, 2007: 316)

In an attempt to deal with this wide variety of audiovisual content, some companies started to employ a method of "shared authorship" (Bernal, 2006: 1), allowing the creative and localization departments to work together. According to the now defunct LISA[1] (2005), localization is the process of "taking a product and making it linguistically and culturally appropriate to the target locale (country/region and language) where it will be

1 Localisation Industry Standards Association.

used and sold". This definition, however, does not take into account the possibility that fan groups may follow very similar procedures without being bound by considerations relating to economic profit or guideline restraints.

In fact, many fans have already started their own processes of localization through the practice of *romhacking*. This term usually relates to modifications and translations of old games such as those for the Super Nintendo Entertainment System, PlayStation One or Sega Mega Drive, but it is possible to find some translations of current top generation videogames on the Web, such as yet another Brazilian fan translation of *Chrono Trigger* for the Nintendo DS.

The decision to use *Chrono Trigger* as the subject of this study was motivated by two main factors: firstly, the game has a large and active fanbase, as can be observed within the forums of *Chronocompendium*[2], and secondly, this game alone has prompted the development of many fan translations. There is also the fact that RPGs in general may well be the most suitable genre for translation studies, as suggested by Mangiron: "RPGs are the most interesting to analyse from a translation studies perspective, as they contain more text and text-types, due to their more elaborate and intricate stories" (2007: 307). The official and fan translations that will be used as the subjects for this paper are indicated in the following table:

Year of Release	Origin	Translators	Language	Type
1995	Nintendo	Ted Woolsey	English	Official
2000	C.B.T	Ear Clock and Tsuhino hime	Portuguese	Fan
2004	IPS Center	Lynx and Mithrandil	Portuguese	Fan
2007	Chronocompendium	Kwhazit	English	Fan
2008	Nintendo	Tom Slattery	English	Official
2010	F.U.R.T[3]	Alizor, Hyllian, Pinguimbozo, lohansan, Gamerulez, Marvin Dalkiri and Merlim	Portuguese	Fan

Table 1: Official and Unofficial Translations of *Chrono Trigger*.

2 The forums can be accessed on line at: <http://www.chronocompendium.com/Forums/index.php?PHPSESSID=55ded7b019470ea3ac1d9d2f1169ce6e;www>.

3 Unified Forum of Translation and Romhacking. The project was open to any member who wanted to collaborate. Online at: <http://www.romhacking.trd.br/index.php?topic=4134.0>.

2. Localization, Video Games and Fan Translations

In many respects, video games are similar to books and movies: all three include a great deal of textual and visual content to be interpreted and at the same try to make us relate to the protagonist of the story. The main difference between video games and books or movies is the way in which video games connect with the player, letting him or her control or direct a large proportion of the actions occurring on screen.

For Bernal (2006: 2), "Video games aim at establishing a different type of relationship with players, that of 'masters of their destiny'". This means that, for videogames, it is necessary to allow the players a feeling of control and freedom in relation to their avatars, although this illusion is limited by the game interface and the programmed variables. The interactive factor between player and game actually becomes an additional challenge for localizers, since they have to direct all their efforts towards the way the game feels in a different language and culture, rather than producing a mere transposition of the source text, as evidenced by Mangiron & O'Hagan:

> The main priority of game localization is to preserve the gameplay experience for the target players, keeping the 'look and feel' of the original. The brief of the localizer is to produce a version that will allow the players to experience the game as if it were originally developed in their own language. (Mangiron & O'Hagan, 2006: 14)

This is the reason why so many wordplays, cultural references and even character names can be expected to be changed. Nintendo of America (NOA), for example, spent six months modifying the game *Animal Crossing (*2001) so that it would become attractive to the American audience (Pham and Sandell, 2003). Among the many observable changes, we can highlight the removal of Japanese public holidays and the addition of American ones such as Thanksgiving, and the changing of hundreds of Japanese names to those more easily recognizable in American culture. The translation/localization of a game is therefore not limited to its textual content.

There are many instances when the legislation or traditions of a certain country do not allow for a game's release without its undergoing a lengthy process of adaptation. It is sometimes necessary to reduce the level of violence or sexual content in a game in order that it can be played in the target country by the same age group for which it was originally produced. According to Dietz, "in some cases, a game cannot be simply translated and released in the target culture, but must first be 'culturalized', i.e., adapted to account for certain cultural conventions and preferences." (2006: 129)

In other cases, however, the game is often too culturally specific, such as in the cases of *Tokimeki Memorial (1994)* and *Legend of Darkness (1998)*. The former is a dating game set in a Japanese high school which was completely changed to suit an American audience. The latter is a Korean MMORPG (Massively Multiplayer Online Role-laying Game) used as the basis for the game *Dark Ages (1999)*, in which the narrative focus was completely changed to one of Celtic mythology when it was localized into English. In such cases, a process of *blending* takes place. According to Thayer and Kolko, "blending involves writing a new game narrative for a target culture, a narrative that will be more comprehensible than the original within that culture and that will hopefully ensure greater sales of the game" (2004: 13).

These characteristics of game translation also led Mangiron and O'Hagan to suggest the term *transcreation* in order to account for such modifications. Because reworking the entire game structure or narrative is often deemed too expensive, most changes are focused on smaller details that might make the game more appealing to its intended audience. In this way, *transcreation* involves a great range of modifications, such as new and originally unintended jokes, the alteration of cultural and geographical references and even the graphical restructuring of some of the game elements or characters:

> No oddities should be present to disturb the interactive game experience, and this is the reason why game localisers are granted *quasi* absolute freedom to modify, omit, and even add any elements which they deem necessary to bring the game closer to the players and to convey the original feel of gameplay. And, in so doing, the traditional concept of fidelity to the original is discarded. In game localisation, transcreation, rather than just translation, takes place. (Mangiron and O'Hagan, 2006: 20)

3. Chrono Trigger

Released in 1995 for the Super Nintendo Entertainment System in Japan, *Chrono Trigger* was a major success amongst RPG fans. Its American version was translated by Ted Woolsey, a translator in charge of many other of Square Enix's successful titles such as *Final Fantasy VI* (1994), *Secret of Mana* (1993) and *Breath of Fire* (1993). Many years later, Woolsey's translation served as the basis for the first two Brazilian fan translations, carried out by the groups *C.B.T* (Brazilian Center of Translations) and *IPS Center*.

The game was produced by a union of talents who became known as the "Dream Team", comprising Hironobu Sakaguchi, the creator of the *Final Fantasy* series; Yuji Horii, a freelance designer and creator of Enix's popular *Dragon Quest* series; and Akira Toriyama, character designer and author of *Dragonball* manga. The resulting game featured an automatic battle system, a time travelling plot focused around avoiding the end of the world in the year 2000, unique characters and the possibility of ten different endings according to the player's choices. The combination of these factors prompted the development of a large fanbase.

The success of the game generated a sequel called *Chrono Cross*, a ported version for Playstation (1999) and another for the portable Nintendo DS (2008). This most recent version involved some graphical restructuring, the addition of three new dungeons and one more ending, but also included a completely new translation. However, why would Square Enix retranslate the game? Why not use Woolsey's translation, since it is not a new game but just an updated version of a previous one?

Some of the underlying reasons may have to do with an interview given by Woolsey in which he stated that memory limitations and time constraints had a strong influence on his work: "Try translating a couple of thousand pages in a month or so (then having to go back and toss out 50% due to memory limitations)" (Woolsey, 1997). This interview is also one of the reasons why fan group *Chronopendium* (2007) decided to retranslate the game. Nintendo, on the other hand, released the new translation with the Nintendo DS game a year later.

In the first official translation of *Chrono Trigger*, a pun involving the names of three famous artists was inserted, as the original pun (based on Japanese food) was considered too obscure for an Anglophone audience.

The fan translator Khwazit, however, decided to transpose the names literally, keeping the same references to food as in the Japanese source text, whereas the new official translation maintained the adaptation from the first translation.

Japanese source text (1995)	*Ted Woolsey* (1995)	*C.compendium* (2007)	*Nintendo* (2008)
ビネガー (binegaa)	Ozzie	Vinnegar	Ozzie
マヨネー (mayonee)	Flea	Mayonnay	Flea
ソイソー (soisoo)	Slash	Soysaw	Slash

Table 2: Enemy names in *Chrono Trigger.*

According to *TVTROPES*[4], a wiki catalog for fiction writing and general pop culture, some fans coined the term *Woolseyism* to describe the way that Woolsey dealt with the many games for which he was responsible: "Woolsey's changes? They *worked*. Some of the lines were so well integrated into the collective consciousness of the game that they have been embraced by the fandom instead of reviled." In order to observe to what extent this generic statement is in fact true, we took some of *IMDB's* (Internet Movie Database) most remarkable *Chrono Trigger* quotes and compared them with the *Chronocompendium* fan translation which, according to its translator, was intended to be the closest possible translation to the Japanese source text, in order to observe the similarities and differences:

> A clearer portrayal of *Chrono Trigger* as intended by its Japanese creators is now available. It is not the opinion of this project that Ted Woolsey's official translation was bad or insufficient in any way – only that some essence of the game was lost or altered, given Nintendo of America's censorship standards and the inability of the game to hold all the original text when translated to English. (Chronocompendium, 2007)

The two tables below contain examples from IMDB[5] and *Chronocompendium*:

4 Available Online at: <http://tvtropes.org/pmwiki/pmwiki.php/Main/Woolseyism>.
5 Available Online at: <http://www.imdb.com/title/tt0210613/quotes>.

Japanese source text (1995)	*Ted Woolsey (1995)*	*C.Compendium (2007)*
魔王「フン、弱い者ほど強がる……。 (Hmph.. a weak one trying to act tough)	[Magus] The weak always strive to be weaker…	[Magus] Hmph, weaker they are, tougher they act….
魔王「歴史が変わるというなら、変われ！ 世界がほろぶというなら、ほろべ！ (If history is to change, let it change, change! If the world is to be destroyed, let it be destroyed)!	[Magus] If history is to change, let it change! If the world is to be destroyed, so be it!	[Magus] If history is to change! If the world is to be destroyed, be destroyed!
魔王「そのために、今この俺が 消えてなくなるというのなら…… それはそれで、おもしろい！！ (If for that I have to disappear… It will be fun anyway)	[Magus] If my fate is to be destroyed… I must simply laugh!!	[Magus] If I am thus to vanish and be no more… That in itself will be interesting!!

Table 3: Magus's lines in Woolsey's Official and *Chronocompendium's* fan translations.

We can easily perceive how the dialogues are modified and recreated by Woolsey, but it is interesting to note that these same modifications are those that drew the most attention from some players and have perhaps made the sentences more captivating, meaning that they appear in IMDB's list of "memorable quotes". By means of adding words or just simplifying Magus' speech, Woolsey attempts to make the character more appealing to the Anglophone audience.

Japanese source text (1995)	*Ted Woolsey (1995)*	*C.Compendium (2007)*
魔王「……	[Magus] ……	[Magus] ……
今度こそ、きさまを倒し…… (This time by defeating you, you bastard….)	It's over for you…	This time, I WILL beat you, you bastard……
わが長き闘いに決着をつけてやる……！ (I will put an end to our long battle.)	Your life ends here!	I will settle my long struggle……!
[Nun]	[Nun]	[Nun]
私達は、世界の平和のために (We are for peace in the world)	We want nothing but world peace…	We are praying for the sake of world peace.
ウフフ……。 (hahaha)	Or a piece of the world, tee, hee…	U, hu, hu……

Table 4: Magus's lines in Woolsey's Official and *Chronocompendium*'s fan translations.

The modifications to Magus' speech in Woolsey's version make him sound more laconic, although this still fits his personality in the game very effectively. We can also observe how a Nun (who is a hidden enemy creature) makes a joke about "world peace", whereas in the fan translation the only hint that something odd is occurring is in the subtle laugh "u, hu, hu…". In the Brazilian Fan translations, this joke is translated almost literally by *C.B.T* and adapted by *IPS Center*, as shown below:

Ted Woolsey (1995)	*C.B.T* (2000)	*IPS Center* (2004)
[Nun]	[Nun]	[Nun]
We want nothing but world peace…	Tudo que nós queremosé a paz no mundo. (all we want is world peace)	Tudo o que nós queremos é a paz… (all we want is the tooth…)
Or a piece of the world, tee, hee…	Ou um pedaço dele… (or a piece of it)	…ta de dentes, hehe!! (paste, hehe!!)

Table 5: The Portrayal of the Nun in the Ted Woolsey and two Brazilian Fan Translations.

The adaptation by *IPS Center* uses the word "paz" (peace) as the first syllable of what later becomes "pasta de dentes" (toothpaste), reworking Woolsey's pun into a joke and still maintaining the feeling that something sinister is happening. The pattern of changes continues with the character Frog, who in the official translation speaks with an antiquated English register, possibly due to his characterization: a heroic and fantastic creature controlled in the medieval period.

The following table compares Woolsey's translation with both the Brazilian fan translations, at the point at which Frog and his King are bidding each other farewell and returning to their own times:

Ted Woolsey (1995)	*C.B.T* (2000)	*IPS Center* (2004)
[Frog] 'Tis a feisty crowd! But they are thine kin,	[Frog] Eles são vossa família, (they are thy family)	[Frog] E pensar que a vossa família (And to think that thy family)
and 'tis of consequence.	e por isso estás aqui. (and that is why you are here)	começou com eles… (started with them)
Queen Leene awaits.	A rainha Lena espera. (Queen Leene awaits.)	Bem majestade, nós devemos ir.
Your Majesty, we too shall take our leave.	Vossa Majestade, eu também (Thy Majesty, I too) terei de me retirar. (shall take my leave)	A Rainha Leene anseia (Queen leene is eager) pelo teu retorno. (for your return)

Table 6: Frog's lines in the Official Translation and two Brazilian Fan Translations.

There is an attempt to reproduce Frog's old-fashioned speech by using the archaic pronoun "vossa" in both fan translations. Additionally, both adaptations move away from a direct translation through either adding or omitting information in order to maintain or increase the atmosphere of the context: in both, the line "tis a feisty crowd!" is omitted, whilst *IPS Center* also adds that "Queen Leene is eager for their return". However, these modifications remain coherent with the entire farewell scene, working just as well as most of Woolsey's changes for transferring meaning from Japanese into English.

For some fans, however, Woolsey also made negative changes to the Japanese script. Frog, for example, was the only character from the game's medieval period to speak in an archaic way, thus generating a small inconsistency in the fictional world. In addition, deadlines, the pressure of being a single translator and the cartridge data capacity available at the time were all factors which influenced Woolsey to reduce as much of the textual volume of the game as possible, using abbreviations or even suppressing lines and dialogues from some characters.

This was another reason why fans gathered around the webpage *Chronocompendium* and retranslated the whole game from Japanese into English. This decision yielded some curious results, such as the removal from Frog's speech of all the quirks and idioms, although it still remained more formal than that of the other characters.

Japanese source text (1995)	*Ted Woolsey (1995)*	*C.Compendium (2007)*
カエル「 (Frog)	[Frog]	[Frog]
にぎやかな連中だな。 (That's a lively bunch.)	'Tis a feisty crowd!	They're a lively bunch.
自分の先祖かもしれな いと思うと ほっといてもいられな いがな…… (I think they may be your own ancestors, so you can't just be rid of them….)	But they are thine kin, and 'tis of consequence.	But they may be your own ancestors, so you can't just be rid of them……
さあ王様、リーネ様が お待ちです。 (So, your Majesty, Leene is waiting.)	Queen Leene awaits. Your Majesty,	Come, your majesty, Leene-sama awaits.
私達も帰りましょう。 (We can go home.)	we too shall take our take our leave	Let us also return.

Table 7: Frog's lines in the Official Translation and the English Fan Translation.

An interesting aspect of the *Chronocompendium* fan translation worth highlighting is the use of the honorific Japanese suffix "sama", which neatly displays the dichotomy between domestication and foreignization (Venuti, 1995). This suffix has no direct equivalent in English, so the translator has to make a choice between foreignizing the translation and moving the reader closer to the Japanese cultural system or following a domesticating approach and erasing all traces of it.

Official translations usually choose the latter option, whereas fan translators may choose either as a dominant approach. The emergence of a fan translated version such as that from *Chronocompendium* may serve a purpose identified by Muscar: "As audiences are exposed to a fan translated version and petition companies for an official release, companies may want to revisit old games…" (2007: 252). She also comments on how these rereleases often result in improved translations, graphics and even additional quests and objectives.

The second official translation for the Nintendo DS (2008) follows this pattern by adding new quests and yet another possible ending. At the same time, it also chooses to maintain Frog's register from the Japanese source text, becoming more literal but not attempting to incorporate foreign elements into the text:

Nintendo DS (2008)
Frog: Exuberant folk. And to think, we may all have descended from them!
Well, Your Majesty, Queen Leene awaits. We, too, should take our leave.

Table 8: Frog's lines in the new official translation.

In the Brazilian Fan Translations, however, the introduction of foreign elements into the text is not necessarily an issue. When Frog is speaking with Chrono's mother, we can observe the difference in the way she is addressed, either as "Madame" or using the foreign contracted term "Milady":

Ted Woolsey (1995)	*C.B.T* (2000)	*IPS Center* (2004)
Frog: M, madam…!	Frog: M, Madame…! (Madam)	Frog: M…milady…[6]
I am NOT a pet!	Eu não sou bichinho! (I'm not a little pet)	Apesar da minha…aparência, não (Despite my appearance, I am
Though I am…odd of figure, I am a	Embora seja…estranho, (Although I am… strange,)	sou uma fera selvagem! (not a savage beast)
master swordsman!	sou um excelente espadachim! (I'm an excellent swordsman)	Sou um mestre espadachim! (I'm a master swordsman)

Table 9: Frog talking with Chrono's mother.

6 It is not uncommon to read or hear the term "milady" in books or in movies dubbed into Brazilian Portuguese.

Further changes can be observed in relation to the names of certain characters. The names of four runners who are competing in a festival were transformed in a way which the player could identify with, echoing local cultural references. *C.B.T* (Brazilian Centre of Translations) uses two very recognizable names: *E.t de Varginha* (a reference to a local myth about an extraterrestrial appearing in the city of Varginha) and *Seu Madruga* (named after a character from the Mexican television sitcom *El Chavo* which was dubbed and aired in Brazil throughout the 1990s and is still shown on Sundays on the TV channel SBT). In contrast, *IPS Center* (the group responsible for the second fan translation) does not create any new puns, although they may have attempted to preserve a connection with the pun of G.I Jogger created by Woolsey, an apparent reference to the American series of action figures "G.I Joe", since in the Japanese source text the character was only referred to as "novice soldier" (ほいほいソルジャーです).

Ted Woolsey (1995)	*C.B.T* (2000)	*IPS Center* (2004)
A. Steel Runner!	A. O Gordinho	A. Corredor de Aço
B. Green Ambler!	B. ET DE VARGINHA	B. Viajante Verde
C. Catalack!	C. Gatuno	C. Gatalack
D. G.I. Jogger!	D. SEU MADRUGA	D. GENERAL LENTO

Table 10: Textual Depictions of the Runners at Leene Square.

Fan translations may therefore follow very different paths with regards to translation choices. Some will attempt to adapt cultural elements, even translating proper nouns, while others will consider that keeping some foreign elements is a way of preserving the game's original feel, as Lynx, the fan responsible for coordinating the second Brazilian fan translation of *Chrono Trigger*, states:

> At first, I was very happy with finally playing a game that I really liked in my own language. On the other hand, I was a bit disappointed with the name changes of the previous translation. Seeing my dear *Schala* written *Sara*, or *Melchior* like *Belquior* and many other changes ended up wasting what could be an excellent translation. Then I used to think to myself: "If I knew how to translate, I could do it better." (Linx – IPS Center translator, 2004) (translated by the author of this paper)

However, in *Chrono Trigger*, one can observe how Brazilian fan translators sometimes move away from the English source text, adding their own interpretations of some fictional elements and characters. This pattern perhaps characterizes an additional cycle of creation, indicating that transcreation (Mangiron and O'Hagan, 2006) can function as a continuous process and is not solely limited to official localization, as in the following example of the English text: "Taban: Yum! Lemonade sure tastes great outdoors!", from the main ending of *Chrono Trigger*.

The table below compares Woolsey's translation of this scene with its portrayal in the Japanese source text and Brazilian fan translations in order to show how the reference to an alcoholic drink varies amongst the translations.

Japanese source text (1995)	*Ted Woolsey* (1995)	*C.B.T* (2000)	*IPS Center* (2004)
	[Taban] Yum!	[Taban] Ummm!	[Taban] Ahh!
タバン「いやー 祭りで飲む酒はうめーぜ！ (Aah, the sake of matsuri[7] is great)	LEMONADE sure tastes great outdoors!	CERVEJINHA da boa!! (Good beer!)	Essa CERVEJINHA é da boa! (This beer is really good!)
タバン「クロノよ。 お前もいっぱいやりなよ。 (Crono, have a drink too)	[Taban] Crono, have a sip!	[Taban] Crono, tome um gole. (Crono, have a sip!)	[Taban] Ei Crono, tome um gole! (Hey Crono, have a sip!)
今日はヒーローなんだからよ！ (because today you're a hero!)	You're the Hero today.	Você é o Herói! (You're the Hero.)	Você merece, rapaz! (You deserve it, boy)

Table 11: Taban and the Drinks Toast in the Brazilian Fan Translations.

The translation of lemonade as "cervejinha" ('little beer') in both translations represents a reading that takes into account not only the authority of the source text but also the context of the scene and the way it fits into Brazilian culture, since it would seem strange to see a middle-aged man like the character Taban making a toast with a cup of lemonade. This is actually

7 A local festival.

quite close to the literal translation of the group *Chronocompendium*, which utilizes the Japanese drink "sake" in the English fan translation, although the Brazilian fan translators were probably not aware of this.

The Japanese term "酒" can be understood either as "sake" or as a reference to alcohol in general. The reason why NOA (Nintendo of America) censored it and utilized "lemonade" instead was probably in order to avoid the game being given a higher age rating by the ESRB (Entertainment Software Rating Board), thus enabling it to be targeted at a larger demographic base of gamers. The version retranslated for Nintendo DS, however, circumvents this problem by suggesting a toast and not indicating what kind of drink Taban is consuming. The latest Brazilian fan translation, created by a group of volunteer fan translators (2010), follows this option literally because in Brazil toasts are also usually done with alcohol.

Nintendo DS (2008)	*C.compendium* (2007)	*F.U.R.T* (2010)
[Taban] Ahh…Nothing better than a toast	[Taban] Yah, the sake you drink	[Taban] Ahh… Nada melhor que um brinde (Ahh… nothing better than a toast)
under the stars!	at a festival is great!	sob as estrelas! (under the stars)
[Taban] Crono, have a sip!	[Taban] Crono.	[Taban] Crono, beba um gole! (Crono, have a sip!)
You're the Hero today.	You have a glass too.	Você é o herói hoje. (You're the hero today)

Table 12: The toast by Taban in the official retranslation and a further two fan translations.

The translation by *F.U.R.T* also modifies dialogues, inserting fragments into characters' speech in a domesticating way. This can be seen in the interaction between Ayla and Chrono's mother:

Japanese source text (1995)	*Nintendo DS* (2008)	*F.U.R.T* (2010)
ジナ「エイラさん……。 (Ayla-san…)	[Mother]: Ayla, dear…	[Mãe] Ayla, querida… (Ayla, dear…)
わかい女の子がそんな姿で あるきまわっちゃダメじゃない。 (A young girl shouldn't walk around looking like that)	A young lady shouldn't dress like that!	Uma mocinha como você não devia vestir essas roupas indecentes! (A young lady like you shouldn't wear those indecent clothes)
エイラ「エイラ　コレ…… (Ayla, this…)	[Ayla] But, Ayla like.	[Ayla] Mas Ayla gostar (But Ayla like).
ジナ「返事は『はい』です！ (The answer is yes!)	[Mother] No buts! The proper response is, |Yes, ma'am!|	[Mãe] Nem mas nem meio mas! A resposta correta é, Sim, madame! (No buts nor half buts! The right answer is, Yes, Madam!)

Table 13: Versions of the dialogue between Ayla and Chrono's mother.

The above examples indicate different approaches of fans and official translators, with the former being more likely to experiment with translation options, either by domesticating or foreignizing the text, and the latter remaining more conservative, bringing foreign elements into the game whilst adopting a predominantly domesticating approach. They also show how fans are steadily escaping from the label of being *consumers* and joining the category of '*prosumers*' (Tapscott & Williams, 2006): i.e. 'producers and consumers'. The nature of this production however is somewhat variable, as indicated by Burn (2006: 88), since fan activities "revere the original text, seeking to remain as true to it as possible" but at the same time "fan work can dramatically alter the original text, adapting it to express the particular interest of the fan or fan group".

A summary of the main approaches used by each of the translations analyzed herein is provided below:

Year	*Origin*	*Main approach(es) adopted*	*Language*	*Translation Type*
1995	Nintendo	Domesticating	English	Official
2000	C.B.T	Domesticating	Portuguese	Fan
2004	IPS Center	Foreignizing/Domesticating	Portuguese	Fan
2007	C.Compendium	Foreignizing	English	Fan
2008	Nintendo	Domesticating	English	Official
2010	F.U.R.T	Domesticating	Portuguese	Fan

Table 14: The main approaches to translation employed by the different translators analysed in this paper.

Conclusions

Through the analysis of fan translations of *Chrono Trigger*, we have shown how fans can either follow very similar procedures to those of official localizers or utilize a "foreignizing" approach. In the case of the former, they adapt the game to reflect their own cultural reality, as indicated by some of the *C.B.T* and *IPS Center* translation choices; in the case of the latter, they keep the game as faithful to the source version as possible, such as was the intention of the *Chronocompendium* translation. We have also highlighted how today's fans are much more than simply passive consumers, being as they are able to organize themselves into online communities and coordinate many different projects, including both translations and re-translations and even the engineering of a fan game.

Additionally, we have established how videogame localization is regulated by a series of specific demands, which may therefore require different translation strategies. The transcreation model, as well as the concept of *blending*, are attempts at understanding how these demands work, although we consider *transcreation* to be the most appropriate approach to help us understand and categorize the variations that occur in both official and fan localizations, since most localizations do not involve the creation of a new narrative.

The articles and theses on game localization thus far researched suggest that official localization tends to focus on reaching the largest demographic base, even if it becomes necessary to censor, modify, add, or remove any aspect of a game for either cultural or legal reasons "so game companies often adapt localized versions to adhere to national rating systems" (Dietz, 2006: 130). We believe that this is the reason why game translators are actually persuaded or allowed to remodel and adapt certain characteristics of a game (even to the point of requesting graphical modifications or plot changes) in order that the game becomes more attractive to its target audience. (Mangiron, 2007: 309).

Fan localizations, on the other hand, usually focus largely on the linguistic aspect of the localization, even though some Rom-hacking projects also include graphical and plot modifications. The choice to use a more "foreignizing" or "domesticating" approach (Venuti, 1995) may depend on the fan group or the individual translator, for there is as yet no evidence that either one of these approaches is favored over the other by fan translators. The translations analyzed, however, indicate that *Chronocompendium* and *IPS Center* focus more on a foreignizing approach, though the latter also employs some domesticating strategies, while *C.B. T* and *F.U.R. T* follow the pattern of official translations by domesticating the majority of the content.

Some issues however, are still yet to be explored. Fan translations need to be further analyzed in comparison to their official counterparts, and reception studies should be carried out to measure the actual impact on the audience of both official and fan translations. As a first step, the quotes from the IMBD were examined, although there is no similar *Chrono Trigger* database for the Brazilian audience. It seems that having a translator work on the transcreational and cultural adaptation of linguistic data makes the game more memorable for the target audience in both official and fan translations, since some puns such as those concerning the three enemies (Ozzie, Flea and Slash) and the four runners from Leene Square (particularly "Seu Madruga" and "E.T de Varginha") can be easily identified with by gamers in the target culture. Nevertheless, confirming this hypothesis would require a more detailed analysis.

Finally, we hope that this paper will serve as a stimulus for more studies concerning localization in general and in particular fan translation. Case studies like this show us how fans are dealing with the games they love whilst at the same reinterpreting and reconstructing them.

References

Bernal, Merino. 2006. 'Localization and the Cultural Concept of Play'. Online at: <www.gamecareerguide.com/features/454/localization_and_the_cultural_.php> (consulted 20.03.2013).

Burn, Andrew. 2006. 'Reworking the text: online fandom' in Diane Carr, et al. (eds). 2006. *Computer Games: Text, Narrative and Play*. Cambridge: Polity Press. 72–87.

C.B.T. 2000. 'Ips archive'. Online at: <http://www.brgames.org/download.php?ipstraduzido=316> (consulted 01.09.2013).

Chronocompendium. 2007. 'Why Retranslate'. Online at: <http://chronocompendium.com/Term/Retranslation.html> (consulted 23.01.2013).

Dietz, Frank. 2006. 'Issues in Localizing Computer Games' in Keiran J Dunne (ed.) 2006. *Perspectives on Localization*. Amsterdam/Philadelphia: John Benjamins. 121–134.

Díaz Cintas, Jorge and Muñoz Sánchez, Pablo. 2006. 'Fansubs: Audiovisual Translation in an Amateur Environment' in *Journal of Specialised Translation* 6. 37–52. Online at: <http://www.jostrans.org/issue06/issue06_toc.php> (consulted 23.01.2013).

Ips Center. 2004. 'Ips archive and Translator notes'. Online at: <http://www.brgames.org/download.php?ipstraduzido=317> (consulted 03.02.2013).

Jenkins, Henry. 2006. *Convergence Culture*. New York: New York University Press.

McKay, Patrick. 2010. 'Culture of the Future: Adapting Copyright Law to Accommodate Fan-Made Derivative Works in the Twenty-First Century'. Online at: <http://ssrn.com/abstract=1728150> (consulted 10.02.2013).

Mangiron, Carme. 2007. 'Video Games Localisation: posing new challenges to the translator' in *PerspectivesStudies In Translatology*, 14 (4). 306–317.

Mangiron, Carme and O'Hagan, Minako. 2006. 'Game localisation: unleashing imagination with 'restricted' translation' in *The Journal of Specialised Translation* 6. 10–21.

Muñoz, Pablo. 2009. 'Videogame Localisation for Fans by Fans: The Case of Romhacking' in *The Journal of Internationalisation and Localisation* 1. 168–185. Online at: <http://mediarumba.com/wp-content/uploads/2010/07/JIAL_2009_1_2009_APA.pdf> (consulted 15.10.2010).

Muscar, Jaime. 2006. 'Winner Is Who – Fair Use and the Online Distribution of Manga and Video Game Fan Translations' in *Vanderbilt Journal of Entertainment and Technology Law 223*. On line at: <http://law.vanderbilt.edu/publications/journal-entertainment-technology-law/archive/download.aspx?id=1694> (consulted 20.10.2010).

Pham, Alex and Sandell, Scott. 2003. 'In Germany, Video Games Showing Frontal Nudity Are OK, but Blood Is Verboten' in *Los Angeles Times* (day month year). On line at: <http://articles.latimes.com/2003/jun/09/business/fi-german9> (consulted 15.10.2010).

Rork Bob. 1999. 'Interview with Ted Woolsey'in *Square-Haven.com*. Online at: <http://chronocompendium.com/Term/Bob_Rork_Woolsey_Interview.html#BR:_What_

were_the_easiest.2Fhar/07est_games_for_you_to_translate.3F_Why.3F> (consulted 05.07.2013).

Thayer, Alexander and Kolko, Beth. 2004. 'Localization of Digital Games: The Process of Blending for the Global Games Market'. Online at: <http://bethkolko.com/includes/pdfs/blendingingglobalgamesmarket.pdf> (consulted 10.05.2013)

Tapscott, Don and Williams, Anthony D. 2006. *Wikinomics: how mass collaboration changes everything*. New York: Portfolio.

Venuti, Lawrence. 1995. *The Translator's Invisibility: A History of Translation*. New York and London: Routledge.

Xiaochun Zhang

Terminology Management in Video Game Localisation

As an emerging form of specialized translation, video game localisation has been the subject of analysis by various disciplines, especially the field of translation studies. Considerable research has examined the nature of video game localisation, following which various translational strategies and skills required for game translators have been proposed. However, rather less attention has been paid to terminology management within this domain, which can potentially resolve many problems arising in the practice of video game localisation. This paper firstly provides a brief introduction to terminology as a discipline. *It then explores the benefits of terminology management in video game localisation before discussing the processes and methods of managing terms in industrial practice. Lastly, it concludes with an* assessment *of and* outlook *for the benefits of establishing an online public terminology database both for the domain of video games and beyond.*

1. Introduction

Video games are rapidly becoming global products that cater for players from various cultural backgrounds and languages, making video game localisation (hereafter referred to as 'game localisation') an increasingly important part of the production chain. With the support of newly evolving technologies, the user interface and the graphical and musical elements of video games are more exquisitely designed. Meanwhile the gameplay and storylines have been created in a more sophisticated way, bringing not only additional enjoyment for players but also extra challenges in the area of game localisation, especially for the translators. In both industry and academia, game localisation has become an important topic in itself, attracting a great deal of interest from theorists and practitioners across various disciplines. There has been much discussion concerning the exclusive features of game localisation and the necessary skills that are required for translators. Some of these challenges are consequences of the rapid formation and development of terminology in the video games' domain, which may perhaps be resolved or reduced by performing effective terminology management. Nevertheless, terminology and terminology management in game localisation have not received much consideration by publishers and localisation service providers (referred to as 'localisation providers' hereafter) within industrial practice. This paper will therefore introduce terminology and the benefits of managing such terminology within the video games industry, as well as discussing the proposed processes and methods of terminology management in the industrial context of game localisation. It will then consider the prospect of establishing a web-based public terminology database for the video games domain and the potential benefits of this both to the industry and academia.

2. The Motivations for Managing Terms in Game Localisation

According to the Oxford English Dictionary[1], a 'term' is defined as "a word or phrase used to describe a thing or to express a concept, especially in a particular kind of language or branch of study". This distinguishes it from a 'word', which is "a single distinct meaningful element of speech or writing, used with others (or sometimes alone) to form a sentence". From these definitions, we can see that a term focuses on a specific concept within a particular domain, whereas a word concentrates on meanings in all types of subject fields. For example, as a word, a 'bug' has several meanings, one of which is a type of small insect; while as a term in the software-specific domain, it indicates an error. Additionally, a term can contain one or more words (Granda and Warburton, 2001: 2), such as 'NPC', which in the game domain refers to 'non player character' and is a term containing three words.

With the development of technology in the late 20th century, the depth of knowledge in various fields has hugely expanded and many new disci-

1 Available at <http://oxforddictionaries.com/>.

plines have been created. This has generated the need for specialized language for each domain to "create and transfer knowledge". "Precision" is the most significant factor in this process, and the only way to guarantee accurate communication within a specific domain and across different linguistic communities is to conduct terminology management (Granda and Warburton, 2001:1). In general, terminology management is "any deliberate manipulation of terminological information" (Wright and Budin, 1997: 1), which, as further explained by Kremer, Kolbe and Brenner (2003: 3), is "the sum of organizational units, processes and instruments that support the creation and management of terms and classifications in a specific subject area". As a discipline, terminology management has a long history and can be traced back to the 18th century (Oeser and Picht, 1997: 341). However, work on terminology did not extend beyond the academic environment into industrial practice until the late 1980s, when the software industry began to mature (Karsch, 2006: 173–174). The expansion of software products into the global market has significantly increased the demand for localisation, which is widely considered to be the main reason for the development of terminology management (Lombard, 2006:149, Corbolante and Irmler, 2001: 516, Warburton, 2001: 677).

According to the now defunct Localisation Industry Standards Association (abbreviated as LISA), localisation is the process of "taking a product and making it linguistically and culturally appropriate to the target locale (country/region and language) where it will be used and sold" (LISA, 1998:3). As further explained by Dunne, localisation involves:

> ...translation of textual content into the language and textual conventions of the target locale; and adaptation of non-textual content (from colours, icons and bitmaps to packaging, form factors, etc.) as well as input, output and delivery mechanisms to take into account the cultural, technical and regulatory requirements of that locale... (2006: 4)

Video games are commonly known as 'interactive entertainment software'. The definition of game localisation can be analogized from the definition of localisation itself, as being the process of modifying an existing video game by translating textual content and adapting non-textual content to make it linguistically, culturally, and technically appropriate to the target locale. In addition, from the perspective of translation studies, a video game is defined by Mangiron as "an interactive multimedia text that combines words, images and sound, and whose main objective is to entertain"

(Mangiron, 2007: 307). Two features of video games are highlighted in both definitions: that they are 'interactive' and 'entertaining'.

Terminology management is considered to be "the backbone of the translation process" (Karsch, 2009: 127) and has been applied to various areas of translation, such as software localisation. However, it has not previously been widely performed in the field of game localisation, which may be due to the general lack of awareness from both publishers and localisation providers of the terminological issues surrounding games and the potential benefits of managing terms. The following section will therefore raise the issues relating to terminology which occur within games and discuss the advantages of performing terminology management in the process of game localisation.

2.1 Terminology management reduces constraints stemming from the source text

The 'interactive' feature of video games has created a number of difficulties in the localisation process with respect to source texts, which rarely occur in conventional text-based translations. Although games resemble books and movies in that they contain a substantial number of audio or visual texts, people relate differently to video games. The relationship between video games and their players is unique, as noted by Bernal (2007). Video games enable the players to be "masters" of their own destiny within the gaming environment, instead of being "spectators" such as they are when watching a movie or TV program (Bernal, 2007: 1). This interactivity leads to the non-linear development of storylines in video games, which affects the work of translation. As Dietz notes, game translators frequently have to deal with source texts that do not always follow a linear flow or have a clear context, unlike in literature or screen translation (2006: 124).

These non-linear and non-contextual texts have produced some problematic issues for game translators and are a major source of mistakes in game localisation. It is not unusual for translators working on game localisation projects to receive source text in an Excel spreadsheet with only one word on each line, as shown below:

	Source-Text	Translation
1	New	
2	Start	
3	Begin	
4	......	

Table 1: Example of a typical game localization Excel spreadsheet.

At first glance, these words appear rather straightforward but they can sometimes be very puzzling for translators. They may question the exact meaning of 'New' in this context. It may potentially be refer to a 'new game' or 'new level' or 'new character' or 'new whatever'. Furthermore, 'New', 'Start', and 'Begin' all potentially mean 'start a new game'. Nevertheless, it is also likely that they each indicate something entirely different in other contexts. For example, 'new' for 'start a game', 'start' for 'start fighting', 'begin' for 'begin a new level'. It is therefore very difficult for translators to make linguistic decisions in the target language when the context of the source text is unclear.

Video game translation is also subject to user interface space limitations, which determines the length and/or number of words that can be used. Therefore, the translators should ideally also know where the translation of 'New', 'Start' and 'Begin' would be used (i.e. in a menu option, as buttons, or as an instruction) and how many words are permitted if they are to appear in the user interface. Although using scrollable windows can free translators from word limitations, as recommended by Mangiron (2007: 308), the translation of these words may vary according to their function and where they appear in a game.

Moreover, the continuous development of video games technology has also introduced many new terms that convey new concepts. A substantial amount of terminological work is thus required to keep track of relevant information on all new terms, including the basics such as their definition. For example, the names of weapons in fantasy-based video games can contain many new concepts, some of which do not even exist in the real world but still require naming, such as the swords 'Andonisus, reaper of souls' and 'ashbringer' in *World of Warcraft* (Blizzard, 2004 – present). Without any terminological information, it may be difficult for anyone to imagine what these terms actually mean, particularly for the translators at the localisation end of the process who may have to create equivalents for these new concepts in the target language.

Furthermore, due to issues of confidentiality within industrial practice, game publishers usually prevent translators (and particularly freelancers) from viewing any detailed information about the games they are being asked to translate. Therefore, the translators cannot actually observe the location of the text on which they are working, or know what the text will look like in a game. The result is that translators frequently have to make a guess about certain ambiguous translatables in the source text based on their own individual experience, which may easily lead to mistakes.

As indicated, the aforementioned issues concerning the source text create substantial difficulties for translators, which may directly lead to unnecessary translational mistakes during the process of game localisation. However, the situation can be transformed by conducting terminology management during the development process. Although terminological work has commonly been considered the duty of the language service vendors, successful terminology management also demands a constant and close cooperation of both the source and target text providers (Esselink, 2000: 398). By managing terms, publishers can to a large extent reduce their inconsistent use in the source text. It is essential to avoid using different terms for one concept or a single term for different concepts, but to stick to one preferred term for each concept (Granda and Warburton, 2001: 5). Returning to the aforementioned example, if 'New', 'Start', and 'Begin' all refer to the concept 'start a new game', it is suggested that one term out of the three is chosen and consistently used throughout the whole development process. If 'Start' is chosen for this concept, 'Start' should not be used for another concept, such as 'start fighting'. A better choice might be for example 'fire' if one begins shooting in the game. Ideally, the consistent use of prominent terms should occur across different document types, across different versions of the same product, across different product lines, and amongst different functional groups (CSOFT, 2010). With proactive terminology management and a systematic terminology database, the publishers would be able to provide translators with the information needed for translation, such as the definition of each term, in which contexts the terms have been used, and whether the translators are subject to any word limitations in translation. With better support from the game publishers regarding the source text, the quality of translation in the localisation process can be significantly enhanced.

2.2 Terminology management improves consistency in the target text

The 'entertaining' feature of video games requires translators to appeal to players through creativity and sometimes humour, which may however increase the inconsistency of the target text. As one of the leading forms of entertainment and an influential new media, video games have always been at the forefront of global popular culture. Game localisation therefore frequently has to deal with culturally bound elements, such as cultural norms and idioms. Translators are required to produce "a version that will allow the players to experience the game as if it were originally developed in their own language and to provide enjoyment equivalent to that felt by the players of the original version" (Mangiron and O'Hagan, 2006: 14–15). However, the handling of various cultural references and different types of humour largely depends on the translators' individual understanding of the source text, their translation competency and previous work experience, all of which can be variable and difficult to standardize.

In addition, game localisation often deals with a large quantity of text, especially in localising MMORGPs (Massively-Multiplayer Online Role-Playing Games). According to the example given by Heimburg, the game *Asheron's Call 2* from Turbine Entertainment (1999) contained 350,000 words in 2006 and yet around 15–20,000 words are added to the game each month (2006: 138). The large volume of text in video games means that game publishers are obliged to assign a localisation task to multiple translators working concurrently on different components. One of the possible consequences of combining individual creativity and teamwork in large-scale multi-lingual localisation projects like video games is the high inconsistency rate in the translations. Without terminology management, inconsistencies can occur very easily. For example, in the English version of *剑侠情缘 3* (*Jiànxiá qíngyuán 3*, also known as *Jx3 online*) (Kingsoft, 2008–present), a Chinese-made online game, the name of one character, '叶姑娘' (yè gūniang) [Miss Ye], has four different translations in the English version that has been published overseas: 'Sister Ye', 'Miss Ye', 'Ms. Ye', and 'Lady Ye'. Amongst all the various localisation tasks required to translate this online game, translating '叶姑娘' (yè gūniang) [Miss Ye] might well require the least creativity, yet the translations vary enormously. Inconsistency can lead to confusion, as players may wonder whether Sister Ye is Miss Ye or a sister of Miss Ye, and query the identity of Lady Ye, and whether or not she is the mother of Miss Ye. This confusion can potentially lead to misunderstanding and thereby provoke in the user frustration and mistrust of the game itself, which may result in losing players to other games available on the global market. Terminology management during the localisation process is therefore vitally important in order to centrally control the consistent translation of key terms and maintain the overall style of background stories and humour, particularly in projects involving several translators.

2.3 Terminology management saves on time and cost

The ultimate goal of terminology management is to increase the speed but reduce the cost of localisation. The issue of time constraints has been frequently discussed or rather complained about by localisation providers in the games industry (Bernal, 2007; Dietz, 2007). In most cases, the localisation process begins when a game is almost finished or in its Beta version, with strict deadlines with extremely short time scales given, particularly since the sim-ship (simultaneous shipment) model is the most popular in the global industry practice. However, as proposed by Wright, if a developer, starts terminology work from as early as the design phase and synchronizes the building of a terminological database with the localisation providers who would prepare the equivalent of the terms in place in the current later localisation phase, a great deal of time could be saved (Wright, 2001: 475–476). Moreover, if the localisation provider identifies any terms that might be problematic later on in the translation process, the developer could still make subsequent adjustments to the source text, which could substantially alleviate costs as it would remove the need for making any changes after the game development phrase had been completed (Lombard, 2006: 162). In addition, both game publishers and localisation providers could benefit from the terminology work by reusing the data for any game updates, or in other similar products. Performing terminology management therefore not only relieves the pressure of time limitations on the localisation provider but also considerably reduces translation costs by recycling terminological data in future projects.

3. The General Procedures of Terminology Management in Game Localisation

The procedures for managing terms in the video game domain can vary across different companies, depending on the budgets available for terminology work, a company's individual needs, their utilisation of different terminological tools and various other factors. The following section introduces general procedures proposed for terminology management and highlights the key issues requiring special attention during terminological work in game localisation.

3.1 The key duties of a terminologist

It would be advisable to have professional terminologists undertake terminological work in game localisation projects working as part of the translation team. However, it would also be possible to assign the job to a senior translator or a project manager within the team, who may have greater experience in game translation. The general duties of a terminologist would include: establishing a terminology database if none exists; researching target text equivalents for source text terms for which no translations are provided in the database; approving and reviewing the accuracy and consistency of the existing terms in the database; providing guidance for translators through constant communication throughout the translation process; and maintaining the termbase with constant updates of the latest terminological information and feedback from translators (Esselink, 2000: 398). Furthermore, the terminologist may also need to contact the software engineers, technical writers, graphic designers and other personnel involved in the game development cycle for additional support or clarification on the definition, context, space limitations and other relevant information concerning terms that present ambiguities.

3.2 Terminology tools

There are various tools which facilitate the process of terminology management catering for different scale localisation projects. One is able to choose a tool according to individual needs and financial budgets. For small scale game localisation, terminological information may be managed with a number of spreadsheets in Microsoft Word or Excel.

Lombard, a Terminology Research Manager from Microsoft Corporation, relates how he personally experienced one instance when 12,000 term entries were managed using this method (2006: 168). However, for professional providers of game localisation, spreadsheets might not be adequate for the efficient management of large-scale multilingual terminological data. In this case, terminology software would be indispensable. There are many different types of terminology software on the market and these generally share typical functions such as terminology storage, the ability to quick search, and automatic insertion in the translation environment or are otherwise compatible with CAT (Computer Assisted Translation) tools (Esselink, 2000: 398). Publishers who need to share terminological information with various users, both internal and external, may require more sophisticated terminology management systems which allow the simultaneous accessibility of multiple users, offer efficient ways for users to give feedback and implement these changes in a timely manner in order to guarantee that all the users are looking at the same database (Karsch, 2006: 175). Moreover, web-based systems have also been developed to provide unified access to terminological information via a single user interface, and this approach has been adopted by an increasing number of software companies (Quasthoff and Wolff, 2003: 142). The tools outlined above could all benefit publishers and localisation providers engaged in the process of game localisation and help them increase the efficiency of extracting, storing, searching and recycling terms within the game domain.

3.3 Terminology collection

The first task when engaged in terminology work is to collect terms. Ideally, game publishers should create for the localisation providers a terminology database of source text terms that have been used in their products. In industrial practice however, publishers do not always provide terminological information. Consequently, localisation agencies often have to construct a terminology database from scratch. General terms in the field of video games (i.e. terms that are not specific to an individual video game product) can be collected before the localisation process begins from game-related documents such as the online help file, the printed manual, marketing materials, and technology reports (Esselink, 2000: 400–401). However, product-specific terms cannot be collected before the tasks are assigned to the localisation providers prior to the commencement of translation. In many cases, the collection of terms continues during the translation process, since translators may well notice terms that have been omitted by terminologists or alternatively raise certain terminological issues they encounter whilst translating that require amendments in the terminology database. Terms can be either collected manually or by running term extraction tools, depending on the needs and preferences in each case. Special attention is also required for terms that do not yet have any equivalents or alternatively have multiple equivalents in the target language, and also for terms for which publishers have preferred to use equivalents in the target language (Zheng, 2007: 1350).

3.4 Terminology reference information

It is of crucial importance to enter additional information besides just the term itself and the target text equivalent in the glossaries. However, the reference information can be categorized in various ways in order to cater for particular individual needs. For game localisation projects, it is suggested that the following information should be included for a project glossary entry:

- Source text.
- Definition.
- Grammatical information: gender, word class, conjugation etc.
- Target text: if it has alternative equivalents, the context for each equivalent is required.
- Context: it is recommended to include the function and location of the term here.
- Other in-game information: e.g. if the term is a weapon used by the main character, XYZ, etc.

- Game name.
- Game version.
- Comments: the terminologist can provide other additional information or guidance for translators, especially concerning terms which may require creativity on the part of the translator e.g. 'XYZ speaks street language'.

The terms that require the greatest consistency in game localisation are very often the names of characters, weapons, items and places, which, taken together, require more attention in terms of the work on terminology. Moreover, slang or regional expressions can also be managed as terms for providing guidance for translators and unifying the register and style of in-game dialogues and background stories. Two examples are provided below which respectively illustrate the reference materials that would be useful for the naming of characters and slang.

Source text	叶姑娘
Definition	character name
Grammatical information	compound noun
Target text	Miss Ye
Context	She appears in the task 'Noble League'.
In-game information	Full name: Ye Haitang
Game Name:	jiànxiá qíngyuán, also known as jx online
Version	3
Comments	Miss Ye is an elegant upper class lady. Be aware of the register and style when translating her dialogues.

Table 2: Translating a character's name from Chinese into English.

Source text	Money
Definition	currency and in-game virtual currency
Grammatical information	noun
Target text 1	金钱值 [Money Value]
Context	Occurs in control menu boxes
Target text 2	钱 [Money]
Context	Occurs in the dialogues of the main characters
Target text 3	钞票 [Paper Money]
Context	Ms. Wang, a non player character (NPC), is from Shanghai. In his dialogue, "money" can be translated as "钞票".
Target text 4	票子 [Cash]
Context	Gou Wa, a non player character (NPC), is a thief in the game. In his dialogue, "money" can be translated as "票子".
In-game information	The monetary unit is Yuan.
Game name:	(for illustration purposes only)
Version	1.0
Comments	Be aware of the various contexts.

Table 3: Translating colloquial and regional expressions from English into Chinese.

3.5 Identification of target language equivalents

The coining of equivalents in the target language may at first appear to be similar to translation. Nonetheless, the perspective of the terminologist differs from that of the translator. The translator aims to choose a correct and appropriate equivalent for a particular context, whilst the terminologist focuses on the identification of terms across many different contexts within a particular domain and aims to create terminological entries for each term, instead of translating the whole text (Karsch, 2009: 131).

The difference between these two perspectives becomes more apparent in the video games domain, since the unique aspects of game localisation require terminologists to deal with various contexts in different locales.

It is important to recognise that when deciding on equivalents for socially and culturally-bound terms, the equivalent of a term can differ from its original concept in the source text. The unique features of game translation, as Mangiron and O'Hagan point out, allow translators a "carte blanche to modify, adapt, and remove any cultural references, puns, as well as jokes that would not work in the target language" (2006: 15). Whilst this releases translators from the strict constraints resulting from a need to be faithful to the source text and thus allowing them to apply their linguistic and cultural knowledge to produce the most appealing translations, it may also lead to the diversified use of terms and the adoption of a different register and style within one game. In addition, in many cases, ideological and cultural issues, such as governmental censorship, different age-rating systems and diversified local customs, also affect the linguistic choices of translators, who frequently have to provide an equivalent in the target text which departs from the original concept in the source text. As a consequence, it is advisable to perform proactive terminology management before the translation begins and to continue updating terminological information throughout the translation process.

Before a video game localisation project begins, terminologists should extract and establish equivalents of the terms which to the best of their knowledge conform to any potential social or cultural constraints. They should conduct research into the regulation of language in video games according to age-rating systems, such as PEGI (Pan European Game Information) in Europe and ESRB (Entertainment Software Rating Board) in Canada and the United States. It would also be useful to review previously translated games with similar content to see how they handle politically or culturally sensitive terms. Game review magazines and online forums for game fans may also be good reference resources for appropriate equivalents. During the translation process, it is also vital for terminologists and translators to maintain constant communication, especially if/when translators disagree on the equivalent of a particular term in different contexts. Again, discussion is required when the translators need to change certain cultural references that require the use of equivalents which express concepts different to the original. For example, in the Chinese version of

World of Warcraft (Blizzard, 2004–present), one of the most popular MMORPGs, the localisation provider changed '盗贼' (dào zéi) [thief] into '隐形者' (yǐng xíng zhě) [invisible man] and '偷窃' (tōu qiè) [steal] was consequently replaced by '搜索' (sōu suǒ) [search] (Zhang, 2008: 47), in order to avoid potential censorship by the Chinese authorities. Both amendments departed from the original concepts expressed in the source text. The terminologists need to notify all the translators involved in order that they apply the new equivalents in their translation. That said, terminologists should also clarify that 'an invisible man' is not 'a thief' in the general context but only in a specific locale, in order that when the term data is reused in other localisation projects in another locale, it does not create confusion and lead to mistakes. Terminologists should continuously update the termbase with the feedback from translators and notify all members of the translation team by adding or amending information in the entries.

3.6 Terminology glossaries

As terminological information accumulates, it may become too disorganised to manage, and would therefore need to be divided into different glossaries, depending on the scale of the project and the needs of the users. There are various options for classifying terms. Esselink suggests that terminology databases for software localisation be separated into three types: operating environment glossaries, client glossaries, and project glossaries that include user interface glossaries (2000: 398). Corbolante and Irmler divide software terminology into computer-specific terms and product-specific terms (2001: 520). With regards to video games, Chandler divides the game components into five assets, namely text assets, art assets, audio assets, cinematic assets, and printed materials (2005: 51). The aforementioned classifications each have different focal points and can be considered as guidance for grouping terminological information. However, from a practical perspective, this article suggests that terms within the video games domain should be classified into three categories: technology-specific terms, genre-specific terms and terms for individual games. Technology-specific terms are terms that are exclusively applied in particular video games platforms (computer, Xbox, Playstation, or Nintendo), such as terms used in the context of hardware and operating systems. Genre-specific terms refer to terms that are used in particular video games genres, such as historical strategy games, war games, sports games or fantasy based games. Terms for individual games are those terms in each particular game which require special attention, such as those which are culturally or politically sensitive. It may be useful to create a glossary of all the translatable items that are subject to space limitations, in order that the translators can be aware of restrictions on the length and number of words whilst carrying out the process of translation. It is also recommended that glossaries are created for all the names of characters, places, tools and weapons for role play games, especially for MMORPGs. However, these glossaries need to be carefully designed according to how they will be practically used in the translation and post-translation review process, taking into account special requirements of the termbase users and the recycling of terminology data for use in future projects.

4. The Establishment of a Web-based Terminology Database for the Video Games Domain

As a consequence of the rapid growth of the games industry, it can be argued that there is an increasing need to share company and industry-specific terminological information between corporations and across business regions. Many domains, such as those of law and medicine, have already established various terminology databases of different scales. A multilingual web-based terminology database for the games domain would also be useful for the industry's future development. Such a termbase could enhance the standardization of terminology usage, which could substantially facilitate the comprehension and exchange of knowledge. It would provide game translators with a tool allowing them to check the terminological information they required and to give feedback on it, which could considerably improve the quality of localisation. It would also contribute to keeping track of terminological development in the games domain, which would benefit all manner of research into games, provide training for future game translators, and may also offer valuable references for other disciplines, such as linguistics and cultural studies. At present, a number of glossaries of game terms have been published on the Internet in different languages by fans and translators[2]. So far, however, no professional online terminology database has been established for this domain. The necessity of such a database has been recognised by scholars in South Korea who have begun building an online dictionary of game terminology (Yun et al., 2003). The establishment of such a large-scale terminology database for the domain of video games requires the cooperation of the publishers, localisation providers and translators working in the games industry. This may seem to be something which is very unlikely to occur immediately. However, it could begin with the performing of terminological management at the levels of both publisher and localisation provider, followed by the establishment of an internal terminology database, with the eventual release of non-protected information into the public domain. Despite the extreme confidentiality of the games industry, a multilingual web-based terminology database would positively benefit both the industry and beyond on a long term basis.

5. Conclusions

The 'interactive' and 'entertaining' features of games have produced several challenges for game localisation, such as dealing with the non-linear and non-contextual texts, space constraints , and the translation of humour and particular cultural references. These challenges could either directly or indirectly lead to translation mistakes or inconsistencies, which may be resolved or reduced by performing effective terminology management. Terminology management is therefore introduced as a method to solve translation issues pertaining to terminology in the process of game localisation. Terminology management can enhance translation quality and accelerate translation speed considerably, by reducing inconsistent use of terms in both the source and target texts, offering support and guidance to facilitate team work in the translation process, as well as providing the possibility of reusing terminological data across different documents and

2 Such as <http://classicgames.about.com/od/glossary/Glossary_Classic_Video_Game_Terms_and_Definitions.htm>, <http://www.frankdietz.com/glossary.htm> and <http://pc.pcgames.com.cn/reviews/zt1/0405/320126.html>.

products. The proposed general procedure of terminology management and the key issues that require special attention provide general guidance for terminological work in game localisation. Successful terminology management requires publishers and localisation providers to work proactively not only at an individual level, but also in a cooperative manner. There is a strong need for a web-based terminology database for the video games domain in order to support information exchange, which may ultimately contribute to the establishment of industrial standards for the games industry. Given that terminology management has not previously been widely applied in the games industry, it is essential to draw attention to this and identify the potential for further improvements. Terminology management lags far behind the current expansion trends within the games industry; the management procedures proposed should be considered in order to keep translation expertise at the level of other aspects of a booming industry.

References

Bernal, Miguel. 2007. *Localization and the Cultural Concept of Play*. Online at: <http://www.localizationworld.com/lwseattle2007/downloads/LocalisationAndGames.pdf> (consulted 28.03.2013).

Chandler, Heather M. 2005. *The Game Localization Handbook,* Hingham, Massachusetts: Charles River Media.

Corbolante, Licia & Irmler, Ulrike. 2001. 'Software Terminology and Localization' in Wright, Sue Ellen & Budin, Gerhard (eds.) *Handbook of Terminology Management*. Amsterdam & Philadelphia: John Benjamins. 516–535.

CSOFT. 2010. *10 Good Reasons to Manage Your Terminology*. Online at: <http://www.csoftintl.com/solutions_localization_terminology_management_10_good_reasons_intro.php> (consulted 28.03.2013).

Dietz, Frank. 2006. 'Issues in localizing computer games' in Dunne, Keiran J. (ed.) *Perspectives on Localization* Amsterdam and Philadelphia: John Benjamins. 121–133.

Dietz, Frank. 2007. *"How Difficult Can That Be?" – The Work of Computer and Video Game Localization*. Online at: <http://www.fti.uab.es/tradumatica/revista/num5/articles/04/04art.htm> (consulted 02.03. 2013).

Dunne, Keiran J. 2006. 'A Copernican Revolution' in Dunne, Keiran J. (ed.) *Perspectives on localization*. Amsterdam: Benjamins. 1–11.

Esselink, Bert. 2000. *A practical guide to localization,* Amsterdam John Benjamins.

Granda, Rita and Warburton, Kara. 2001. 'Terminology Management as Data Management' in Stewart, Darlene A. and Johnson, J. Howard, (eds). In proceedings at CASCON '01

Conference of the Centre for Advanced Studies on Collaborative research, 2001. Toronto, Ontario, Canada: IBM Press.

Heimburg, Eric. 2006. 'Localizing MMORPGs' in Dunne, Keiran J. (ed.) *Perspectives on Localization.* Amsterdam and Philadelphia: John Benjamins. 135–154.

Karsch, Barbara Inge. 2006. 'Terminology Workflow in the Localization Process' in Dunne, Keiran J. (ed.) *Perspectives on localization.* Amsterdam: Benjamins. 173–191.

Karsch, Barbara Inge. 2009. 'Profile of a Terminologist in Localization Environments' in *The Journal of Internationalisation and Localisation* (1): 122–148.

Kremer, Stefan, Kolbe, Lutz and Brenner, Walter. 2003. 'Do You Know Your Terms? – A Procedure Model for Terminology Management'. In proceedings at ECIS 2002 Proceedings, 2003.

LISA. 1998. *Report of the Education Initiative Taskforce of the Localization Industries Standards Association.* Online at: <http://leit.lisa.org/pubs/public.pdf> (consulted 28.03.2011).

Lombard, Robin. 2006. 'A Practical Case for Managing Souce-language Terminology' in Dunne, Keiran J. (ed.) *Perspectives on localization.* Amsterdam: Benjamins. 149–171.

Mangiron, Carmen. 2007. 'Video Games Localisation: Posing New Challenges to the Translator' in *Perspectives: Studies in Translatology* 4 (14): 306–323.

Mangiron, Carmen and O'Hagan, Minako. 2006. 'Game Localisation: Unleashing Imagination with "Restricted" Translation' in *The Journal of Specialised Translation* (6): 10–21.

Oeser, Erhard and Picht, Heribert. 1997. 'Terminologieforschung in Europa: ein historischer Überblick' in Hoffmann, Lothar, Kalverkämper, Hartwig & Wiegand, Herbert Ernst (eds.) *Fachsprachen: Languages for Special Purposes.* Berlin: Walter de Gruyter. 341–347.

Quasthoff, Uwe and Wolff, Christian. 2003. 'Web Services in Language Technology and Terminology Management' in *"Multilingual Knowledge and Technology Transfer", proceedings of the 6th TAMA (Terminology in Advanced Management Applications) conference in South Afica:* 142–146.

Warburton, Kara. 2001. 'Globalization and Terminology Management' in Wright, Sue Ellen & Budin, Gerhard (eds.) *Handbook of Terminology Management.* Amsterdam & Philadelphia: J. Benjamins. 677–696.

Wright, Sue Ellen. 2001. 'Terminology as an Organizational Principle in CIM Environments' in Wright, Sue Ellen & Budin, Gerhard (eds.) *Handbook of Terminology Management.* Amsterdam & Philadelphia: John Benjamins. 467–479.

Wright, Sue Ellen and Budin, Gerhard. 1997. 'Introduction' in Wright, Sue Ellen & Budin, Gerhard (eds.) *Handbook of Terminology Management.* Amsterdam & Philadelphia: John Benjamins. 1–10.

Yun, Tae Soo,et al. 2003. 'The Implementation of the Electronic Game Terminology Dictionary' in Sembok, Tengku Mohd Tengku, Zaman, Halimah Badioze, Chen, Hsinchun, Urs, Shalini R. & Myaeng, Sung Hyon (eds.) *Lecture Notes in Computer Science, Digital Libraries: Technology and Management of Indigenous Knowledge for Global Access.* Berlin & Heidelberg: Springer. 480–488.

Zhang, Xiaochun. 2008. '"Harmonious" games localization for China' in *MultiLingual Computing* 99: 47–50.

Zheng, Guozheng. 2007. 'Terminology Management and Term Discussion in Software Localization' in *Computer Knowledge and Technology* 17: 1350.

Stephen Mandiberg

Games, Localization, and Diaspora

The focus of this paper is diasporas and the diasporic game. The paper elaborates as to how games localized for the Chinese locale are inaccessible to people from across the Chinese Diaspora and discusses the various ways that different languages are implemented within games. It also explores the issue of how a game could be created to both represent and support the multiplicity of the languages of the Chinese Diaspora.

1. Introduction

Localization is the process of translating games for distribution in new cultural contexts, and includes linguistic, audio, visual and gameplay[1] alterations. It is espoused by the video game industry as being a more particular, more accessible form of translation catering to the individual player, and because it helps games to cross easily over frontiers and borders, and between people and places. And yet this adaptability is stymied when localization is considered in relation to unusual populaces in alternate locations. Localization seeks to render games accessible, yet it does so by translating towards generic cultural attributes of groups instead of towards the complexities of real groups that do not fit generic moulds and stereotypes. Diasporas are types of populace that reveal this problem of game localization, as the

1 The author refers to localization as enabling linguistic, audio, graphical, and gameplay alterations. Localization companies do not necessarily have control over graphical and gameplay elements, which can include changing character costumes and removing culturally 'inappropriate' mini-games or side-quests such as within Yakuza 3 (Fresh, 2010). These alterations are often performed by the developers, but they are done in order to facilitate the *same* goal of accessibility and global distribution. Thus, our argument labels both translational and programing sides 'localization.'

heterogeneous quality of a diaspora is at odds with localization. These problems are particularly visible in the Chinese Diaspora. People in the various locales of the Chinese Diaspora (those who identify themselves as Chinese, but cannot or do not identify as Mainland Chinese as they live in Hong Kong, Taiwan, Australia, the United States, or many other places) are not the same as their mainland counterparts, but are forced into particular locale based-identities through game localization.

This paper begins by defining 'diaspora' and 'the diasporic text'. It then elaborates as to how games localized for the Chinese locale are inaccessible to people from across the Chinese Diaspora. Next, the paper discusses the various ways that different languages are implemented within games. Finally, it returns to the issue of how a game could be created to both represent and support the multiplicity of the languages of the Chinese Diaspora. At issue throughout is a very real problem of accessibility, where the dominant practice of localization cannot address certain populations despite (or perhaps because of) its claims to fit to a locale. The focus of this paper is diasporas and the diasporic game, but the problems that are introduced involving localization are systemic. As more people enter a state of global mobility and identities become less tied to nation-states and locales, what it means to be 'in diaspora' or 'in a locale' alters. As more people become 'unusual' populaces, changes in the practice of localization become all the more necessary. Game localization practice should adapt to fit these alternate identities immediately.

2. Diaspora

Diaspora has been variously theorized as being exile from a homeland with the dream of return; as the acceptance of a lack of return coupled with the maintenance of a frictional relationship between a mythical homeland and one's current location; and as the contemporary practice of transnational migratory living in the global community (Clifford, 1997; Gilroy, 1993; Liao, in press; Tölöyan, 1996). Following on from the second and third views, the author sees diaspora as a way of life, and identity as frictionally derived from this way of life. It is this pervasiveness of diasporic identities that should compel the game localization industry to change in order to facilitate translational practices more oriented towards the idea of diaspora. While the discourse of diaspora takes a broad range of forms, particularly relevant towards understanding the possibilities of a relationship between video games and the Chinese Diaspora are the contention by Brent Hayes Edwards (2001) that diaspora must be understood through difference, and Shih Shu-mei's realignment of Chinese Diaspora towards a discussion of the linguistic entanglement between people in diaspora. (2007)

Following the discussion by Gilroy (1993) of routes and the Black Atlantic, Edwards argues that there must be an acknowledgement of the lack of return in diasporic populaces. This lack of return leads Edwards to link diaspora to the concept of *"décalage"*, which he defines etymologically as "the removal of such an added prop or wedge… [to reestablish]… a prior unevenness or diversity; it alludes to the taking away of something that was added in the first place, something artificial, a stone or piece of wood that served to fill some gap or to rectify some imbalance". (Edwards, 2001: 65) For Edwards, it is necessary to remove the prop of 'similarity' from discussions of diaspora, because it is 'difference' that defines, produces and helps groups in/of diaspora. Edwards, like many post-structuralist theorists, elaborates a difficult and painful productivity,[2] but a productivity that is absolutely necessary in a world that slowly recognizes the problems with modernity, linear progress, and Capitalist production.

In this context, difference is both productive and necessary to avoid the essentialism rooted within many theories of Chinese Diaspora that seek to locate an essential 'Chineseness' with a shared past. The attempt by Mainland China to assert and attach a national identity onto Overseas Chinese populations is perhaps the most visible and problematic in terms of global power struggles and individual self-identification. However, Tu Wei-ming's (1991) reversal of this logic is just the functional equivalent. The switch by Tu of the center for the periphery, of Mainland China for Taiwan/Hong Kong/etc., does little more than alter the Chinese essence that all others must follow. Like the move of Edwards away from 'similarity,' Wang (1999) locates the differences of the plural Chinese *diasporas*, but still ultimately links the different groupings under one essence.

2 Throughout this paper we use what might be seen as negative terminology such as friction, complicated, difficult, and difference. We take this from the poststructuralist methodology. None of these are negative terms. Rather, they are as complicated and multivalent as life as we know it.

Shih avoids focusing on the essentialness of Chineseness by moving the discussion away from the ethnic *Han*-centric Diaspora towards a study of the Sinophone.[3] The Sinophone is the linguistic link between disparate communities, meaning that there are rough ties between people who use a similar linguistic base. Like the Anglophone and Francophone, the Sinophone points to Imperialist pasts with vestiges in the present; linguistic ties of the present connect us back to a history of Imperialism. Whilst it is a unit that links multiple communities together, the Sinophone is not an essential quality. Shih first notes that "Diaspora has an end date," indicating that there is an end to any identity-driven power link between nations, people and homes. (Shih, 2007: 185) She also notes that "The Linguistic community is a community of change and an open community", which alters the shared quality within the diasporic group away from an essentialness to a forever-changing linguistic tie where no language ever stays the same. (Shih, 2007: 185) These two key points move the discussion away from centuries-old arguments over the essence of 'Chineseness' by decoupling home and origin so that mobility can be politicized, and transcend national borders. Sinophone thus requires "a critical position that does not succumb to the nationalist and imperialist pressures". (Shih, 2007: 190) Although there are strong pressures to submit to the currently dominant nationalist or imperialist means of understanding the world and how people exist within it, Shih, like many working within cultural studies, helps us to understand that whilst it might be difficult, a critical and adaptable position is key to understanding diaspora as well as other global flows (Appadurai, 1996).

These arguments point to the problem inherent in believing that a diasporic community is essentially linked to historic origin or new home locales of identity formation. Being in a state of diaspora is about living with the friction of these and other poles of identity. Diasporas are not easy, but they are a key element to the frictional global mobility in which people are increasingly living in the present era. (Appadurai, 1996; Urry, 2007). Unfortunately, localized games force the diasporic group towards either origin or destination, and define the group as an essentially similar one. In contrast to the reduction to similarity within typical localizations,

3 Similar to the way in which Anglophone and Francophone studies seek to understand the way that English and French imperialist histories have a lasting linguistic mark on populaces around the globe, the emergent field of Sinophone studies seeks to understand China's history as a cultural and linguistic empire (Shih et al., forthcoming).

diasporic localizations should bring out the friction of life in diaspora, referring not simply to the "flows" but also to the stoppages.

The author approaches diaspora by theorizing how one could make a diasporic game. This is not to be confused with a game *about* diaspora. There are games where the narrative involves exiled characters, or traveling between cultural and national entities; these games depict diaspora, but are not necessarily themselves diasporic. Whilst localization pushes the user into a specified identity (a locale), a diasporic game would allow the productive play of difference and multiplicity. Instead of forcing the player into a pre-determined mold, the diasporic game would simply allow the user to identify his or her own route, allegiance and language from a host of possibilities. In concrete terms, this would mean that a Hong Kong-Chinese American dwelling and playing games within the United States should be given the ability to practice his or her identity in its multiplicity through non-reductively localized narratives, and through the ability to choose between Mandarin, Cantonese or a plethora of other topolects (regional spoken dialects) and languages in addition to English, French and Spanish. In contrast to this ideal of a diasporic game, the current system of localization hides the hybridity and multiplicity of games under a guise of localized specificity. With localization, people are reduced to falling into an essentialized group, despite the very real differences between them and the other members. If we follow Edwards' claim of the productivity of difference (2001) , then this reduction is not productive, even if it makes the games industry money.

Like Mirzoeff's diasporic *visual image*, which "can create multiple visual and intellectual associations both within and beyond the intent of the producer of that image" (Mirzoeff, 1999: 7), the diasporic *game* should involve the productive difference between a multiplicity of play points and players from different particular spaces. For Mirzoeff, the place of vision is relocated to a third space, a 'between' space incorporating center and periphery, nation and diaspora. Mirzoeff's third space, which he takes from Homi Bhabha (1994), is not necessarily a fun and comfortable one. However, such third spaces are a common part of the current world system, and are productive in terms of how we understand through them. The author believes the creation of a diasporic game (as a third space) could be worthwhile, as it can help represent the complicated relationships between people and groups, and because it can uncover new possibilities of localization practice.

3. Diasporic Texts

The following section focuses on how literary and cinematic texts have functioned diasporically. Through understanding how other media have intersected with diaspora, we can better understand the possibilities of games as diasporic texts. The two diasporic texts that this paper focuses on are the 2007 novel *'The Brief and Wondrous Life of Oscar Wao'* by Junot Diaz, and Ang Lee's 2000 film *'Crouching Tiger, Hidden Dragon'*. These texts are diasporic because their authors represent complicated relationships with various tangled places and people, but also achieve this complexity through formal manipulation of the respective media. Both content and form are manipulated within the diasporic text.

In *The Brief and Wondrous Life of Oscar Wao*, Diaz utilizes English, Spanish, 'Spanglish' (a linguistic combination of Spanish and English) and his own and his characters' mixing of cultures to bring out the problems and inevitability of, and ultimately come to terms with, the essential hybridity of diaspora. The novel tells the story of Oscar de Leon, an overweight, Dominican-American sci-fi nerd as he grows up in New Jersey, goes to school at Rutgers and eventually travels back to the Dominican Republic, from where his family fled in the middle part of the 20th century. Between the chapters relating Oscar's story, the novel moves forwards and backwards through time, focusing on Oscar's sister, mother and grandfather coming to terms with their diasporic lives as they move between the Dominican Republic and the United States. The story highlights the political turmoil during the 20th century that results partly from the international flows of people and culture, but directly relates to the spread of global capitalism in modernity. However, the association of *Oscar Wao* with diaspora is not simply to do with the narrative and socio-historical context. Rather, there is a temporal, spatial, national, and linguistic mixing that occurs throughout the story and with the words themselves. Oscar comes to terms with his background, masculinity, nationality, and hybrid identity at the same time as the reader comes to terms with the greater diaspora between the Dominican Republic and the United States, and the mixture of Spanish and English that has created the much-used Spanglish in the novel. Key in terms of constructive difference, there is no ultimate linkage between the various moments in the book, and it is only through the non-linear temporal jumps between different generations and countries that the story reaches its conclusion. The non-chronological route provides a reading multiplicity that, combined with the novel's hybrid content and form, makes the text diasporic. While *Oscar Wao* tells the story of a hybrid, diasporic family, of far greater importance is the fact that it interacts with the diaspora as a multilingual, polyphonic novel sold within the context of the early 21st century United States. As a diasporic text, *Oscar Wao* pushes against the chauvinism of the monolingual English-speaking American audience. It also speaks to a global culture of increasing hybridity and embraces the difficult task of highlighting the difference of diasporas through both form and content. It can therefore serve as a useful example of what a diasporic game could do in the same socio-historic context.

Crouching Tiger Hidden Dragon juxtaposes different spoken topolects (regional dialects) and regional actors in order to break with local homogenized standards, thereby highlighting the differences between the actors and actresses in the film, and the friction between geographical locations of the Chinese Diaspora. The standard practice of post-production overdubbing with a standardizing, homogenous Mandarin did not occur in the film, despite the fact that two of the main actors are native Cantonese speakers and all the actors have disparate accents – Zhang Ziyi is from mainland China, Chang Chen from Taiwan, Chow Yun-fat from Hong Kong, and Michelle Yeoh from Malaysia. Through the different bodies, voices and languages of the characters, Ang Lee portrays a vision of the Chinese Diaspora (of which he, as an NYU trained Taiwanese filmmaker, is a member) that has been retrofitted into the format of a *wuxia* film, a genre often used for more nationalistic sympathies. As Shih argues in the introduction to *Visuality and Identity* (2007: 3), whereas the *wuxia* genre typically highlights the importance and dominance of an essential *Han*-centric Mainland Chinese culture, *Crouching Tiger* turns this on its head by highlighting the differences within the genre and the importance of the varied Chinese diasporas that are linked through language, albeit often in an uneven way. *Crouching Tiger* uses audio and visual registers to highlight difference and force the viewer to acknowledge the multiplicity within diasporic populaces (or, as Shih articulates, within the people of the Sinophone). Diaspora is not necessarily linked to the old home (Mainland China), nor is it linked to the new home (be it Taiwan, Hong Kong, Malaysia, or the United States). Rather, diaspora is an important element of the current world system that does not fit within the current definitions of identity through national or cultural association.

Finally, it is necessary to point out that both texts are diasporic through the specificity of their respective forms. The novel uses the written word, and the readers' paths through it, to create an idea of hybridity. By mixing English and Spanish on the textual page, the form of the book itself becomes diasporic. Similarly, the film uses generic conventions and accented dialogue to render visible ideas of the Chinese Diaspora. A game could use the specific elements of play, digital narrative and code to bring out the idea of diaspora. The narrative and the way the player traverses it can certainly encourage a diasporic reading, but the material form of the game code itself can facilitate a player to interact with diaspora. By this, the author means that the way in which the digital artifact is constructed and functions, how it loads up particular assets that can be organized in ordered file structures and how it is played on particular hardware all hold amazing possibilities for enabling a diasporic experience.

4. The Language of Games in China

To move the discussion from diaspora in a broader sense to implementing a diasporic game for the Chinese Diaspora it is necessary to understand the state of language in games, and particularly of games in China. Language resides on multiple semiotic layers. While this statement might cause great consternation to a non-academic audience, its incorporation within video games is key to the differentiation between translation and localization. Within games there is the visual linguistic register that a player encounters, involving both diegetic objects such as in-game signs and billboards, and nondiegetic elements such as subtitles, loading screens and menus.[4] Language is also the spoken dialogue in both the main game and in any cut scenes, and can be in nondiegetic soundtracks. However, language in games also includes other sign systems including visual iconographic material (character representation and programmed world) and the concept of play which, while theoretically universal, is always located

4 The terms diegetic and nondiegetic are utilized within literary and film theory to indicate inside and outside the story world. Diegetic indicates within the story world; nondiegetic indicates not in the story world.

in particular spaces and at particular times (Flanagan, 2007; Galloway, 2006). The many layers of language are one of the justifications given for full localization practices as opposed to traditional translational practices, or "non" localizations. (Chandler, 2005) According to this understanding, localization is a more extensive form of translation as it attempts to translate more semiotic layers of the game than 'normal' translation. However, two issues with localization can be seen within the Chinese context. The first, is whether localization actually occurs. A lack of localization is particularly visible in the Chinese context, where many games played are often illegally imported or simply pirated untranslated Japanese and English localizations, or alternatively are games for handheld devices and the Internet where certain registers such as spoken language are not implemented. The second issue is that even with a full localization, games are not rendered accessible to all, something which can be observed with games that are localized into Chinese/Mainland China, but not necessarily to the disparate people of the Chinese Diaspora. Both of these issues will be more fully explained in the following two paragraphs, and the implications identified will form a thread through the rest of the paper.

Game consoles have been banned in China since the year 2000. (Ashcraft, 2010a; TradingMarkets, 2010) While this does not block black and grey market practices (ESA, 2011), it does limit the availability of newer, more technologically advanced games and game systems unless they are illegally imported or pirated (ESA, 2011). This also means that cheaper, easily smuggled handheld systems with less storage capacity and processing power have been more prevalent over the past ten years. While graphics are often a primary element reduced with limited capabilities, more important for the argument put forward in this paper is that spoken dialogue has largely been missing in Chinese games due to the limited capabilities of portable systems. Spoken dialogue is also strategically omitted due to the different topolects within the mainland and diasporic Chinese contexts. Typically, Chinese games utilize written language that can be read by different regional audiences, despite the particulars of different regional topolects. Thus, the linguistic option given within most games has been, and still is, a choice between simplified and traditional scripts. These correspond to traditional and post-Mao revised characters, but they do not link to a particular locale. This legal and technological situation previously obscured the issues of localization and language within Chinese gaming. However, the dual-script solution is increasingly unsatisfactory, as spoken dialogue rather than just text has become a common form of language in games even on more technologically limited platforms. Where ten years ago handheld systems could not deal with voice and video, even the more limited systems now can. The PlayStation Portable and Nintendo DS systems feature spoken dialogue, and these systems are increasingly pushing against the 2000 console ban due to grey market sales. Thus, issues of language are becoming less easy to ignore within the Chinese linguistic and industrial contexts.

Faced with the increase in voiced dialogue, the Chinese games industry will likely follow convention and overdub games with Mandarin. While in most places this convention goes unquestioned, it becomes visibly problematic when coupled with the Chinese national practice of dubbing cinema in Mandarin to obscure diasporic and local differences (Shih, 2007: 2). Dubbing erases any heterogeneous linguistic elements within a game, making it even harder to see the heterogeneous qualities of groups in reality.[5] Dubbing creates a Chinese 'locale,' where all players are forced into interacting with Mandarin regardless of their geographic identity or cultural differences. While this matches standard localization practice, it renders these games inaccessible to members of the Chinese Diaspora who reject the dominant, *Han*-centric, Mandarin-speaking understanding of their diaspora. The goal of localization to render a game accessible fails when we consider real people and real places, instead of idealized 'locales'. However, localization itself is far from a failed endeavor. Rather, we can use this situation to envision new possibilities for localization to render games accessible to alternate player groups, through practices such as the creation of diasporic games.

5 An interesting rejoinder to this is work done on the mainland Chinese World of Warcraft player base migrations to Taiwan servers. These two player bases find linguistic ways to differentiate themselves. Certain interactions are frictional, but considering the deep political rift between China and Taiwan the mere interaction can sometimes be considered better than nothing (Lin and Sun, 2010).

5. Language in Gaming: Monolingual, Multilingual, and Omnilingual Styles

In order to determine new possibilities of localization, there now follows a discussion as to how languages are currently implemented in conjunction with the materiality of games. This section is primarily technical, but the context of the Chinese Diaspora is returned to in the concluding section of the paper. There, the more technical section is combined with the previous context of games in China in order to discuss how a diasporic game could be implemented. The basic styles of how language exists in gaming are referred to as monolingual and multilingual. The monolingual style of language implementation has a single language environment; the multilingual style creates an environment with more than one language. The multilingual style leads towards what the author calls the 'omnilingual' style, the scenario when a diasporic mode of language integration (i.e. a game in diaspora) could be created.

Monolingual Language Implementation

The most common type of language integration and the base version referred to here is monolingual localization. There are three variations: a) the pure monolingual localization, b) the multilingual installation that installs a monolingual localization, and c) the multilingual disk that loads a monolingual localization. All three variants have *only one accessible language*, as there is only one language available, installed or loaded.

The classic example of pure monolingual localization is the console game designed for entry into one marketplace with only one language available on the disk. The author's copy of the PlayStation 2 title *Ratchet and Clank* (Insomnia Games, 2002) has one language (Japanese) and the DVD is encoded for region 2.[6] There is, of course, an equally monolingual English original/localization on a region 1 encoded DVD.[7] The two

6 In the following section the author utilizes games that he has played, which happened to be nearby on his shelf.

7 The author uses both the terms 'original' and 'localization' here. While it is a ripe area of research, the author is not making an argument regarding version authenticity.

games are distinct products sold to different regions with different regional encoding and they each include only one language. For the localization industry, selling game localizations with a single language and the ability to play on only particular regional hardware is good business. To players raised on the idea that products are created just for them, monolingual localizations appear to be specifically tailored experiences designed to fit their specific locale in their single language. Discursively, this is how we understand games should be: fluid, smooth and fun.

The second variation is the multilingual installation that reads the operating system and installs the corresponding monolingual language environment. This is common with PC games that are marketed to multiple linguistic communities simultaneously. An example is Blizzard's European version of *World of Warcraft* (2004); the installation disks include five localized game environments (English, French, Spanish, German, and Russian). The installation process installs one language, depending on the location of the realm and the OS of the computer.[8] Thus, a Parisian with a French OS will normally be stuck with a French localization. This is a problem for example for an Algerian-French Arabic speaker living in Paris. It is also a problem for a German living in Paris who would prefer to play on a German server. However, the very *possibility* of switching is itself a step forward – Blizzard's installer for *Diablo II: Lord of Destruction* (2000) locked the user into a monolingual environment with the only way of switching languages involving finding complex instructions on the Internet and essentially hacking through the language change (Kilinc et al., 2006). This is above the ability of many players, and certainly the very possibility of doing this was below most users' radar.

The third variation – a multilingual disk that automatically loads a monolingual version – is Microsoft's required standard for their Xbox and Xbox 360 consoles.[9] All localized versions exist on a single disk and are

8 Blizzard has created separate manually installable Language Clients that allow the user to switch between linguistic environments. However, installation requires these to be found and manually installed, something beyond the capability of many users who may not even know the option exists. See: <http://blue.mmo-champion.com/topic/138460/can-i-play-wow-in-spanish/>.

9 This is also the method of many iPhone games, which often have multiple languages available, but the other languages are hidden unless the user switches the iPhone OS's language.

simply read upon loading the game.[10] The game reads the Xbox's OS and loads the corresponding single language. In order to change the game's language, the user must change their 'location and language' in the Xbox preferences menu. However, the average user is unlikely to figure out that the game software contains multiple languages, and is even less likely to work out how to switch the language as the option is buried so deep in the system OS menu. The user would be more likely to discover how to switch the language if the different available languages were visible in the game's option menu, as is the case with DVDs.

This third monolingual implementation is currently the most common. The 'original' is translated into various 'localized versions' that are then combined onto one disk and simultaneously shipped globally – a process called 'simship.' While the Xbox 360 version of the monolingual localization has multiple languages, by hiding the multiplicity of languages it cannot rightly claim to be a multilingual edition. Such monolingual games are by no means bad; for most users monolingualism is one less distraction; the monolingual localized game seems as if it were customized specifically for them; it feels right that way; it sells well. This is the evolution of the original monolingual language implementation and is just as fluid, smooth and fun. However, monolingual games block out the foreign from gaming environments through domesticating translation practices, and force users into a singular language identity. Diasporic users are locked into a singular language that forces them into singular, nationally determined language identities. This is notably a problem for diasporic users, but in a global/hybrid world the assumption of a nation linked to a singular language is quite problematic for all. In terms of a game's accessibility to alternate populations, the monolingual localization fails.

Multilingual Language Implementation

The above monolingual language implementations are considered standard practice for developers and localizers. What the author refers to as a 'multilingual implementation' is less common, but not altogether rare. Unlike monolingual language implementation, multilingual language

10 While the goal is to include all localized versions on one disk, certain versions may need to be released individually due to censorship.

implementation informs the user that there is more than one language available, and makes it easy to switch between languages. Such a multilingual style is not very different from a film on a DVD with multiple language and subtitle tracks.

It must be noted that multilingual language implementation, and the omnilingual language implementation discussed in the next section, are both referred to in the industry as *partial localization*. With partial localization, a game's in-game linguistic elements are translated, spoken audio remains in the foreign language and is subtitled, and further alterations to graphical and gameplay assets are generally not implemented. At present, within the localization community, this style of localization is considered subpar as it does not create a fully tailored environment for its audience. It does, however, allow disparate groups to experience playing what is recognizably the same game (as they can observe and engage with many of the same assets) at a lesser cost than for full localization. Additionally, if given the option to choose their own language, the multilingual implementation can allow a particular audience to choose their own entrance point into the text which is not based on the forced 'locale' that comes with regional encoding and sales.

One variation of the multilingual language implementation is a title screen switch. Ubisoft and its affiliate company Gameloft regularly use this option. Upon loading Ubisoft's *Might and Magic: Clash of Heroes* (2009) – a Nintendo DS title developed by Capybara Games – the user is confronted with three flags representing English, French and Spanish.[11] The user picks one of the three and the game then loads in this language. Once the game is loaded, the session is locked into that language. The user can switch languages by saving, returning to the title screen, switching, and then reloading the saved game file. While the process of switching languages takes a certain amount of time, the user is aware that he or she can switch languages because he or she is forced to determine the language every time he or she turns the game on. Thus, the fact that the game has those three languages is very apparent. One of the benefits of this type of multilingual switch is that it allows certain localized graphical and gameplay assets to be loaded, meaning that the localization need not stop at simply

11 The choice of French, English and Spanish, and the implementation of multilingual instead of monolingual language implementation, is highly relevant for Ubisoft as a company based in France with affiliations in Canada and the United States.

altering dialogue, but rather may change graphics and sounds as desired. For example, there might be three different loading screens and three different sets of character graphics, which are loaded depending on the initial language choice. However, this method of localization does not facilitate censorship, which requires the elimination of particular elements entirely. Even if something were deemed unacceptable and removed from one of the linguistic implementations of the multilingual language implementation, the existence of that same uncensored element in the other selectable languages would cause problems.

A second variation of the multilingual switch is the option menu switch. This method allows the user to change the game's language through the option menu by pausing at any time, switching, and then resuming gameplay. Square-Enix's iPhone title *Chaos Rings* (2010) allows this type of switching, meaning that at any time when he or she is not directly engaged in battle or dialogue, the user can switch between English and Japanese.[12] The option menu switch grants the user the ability to easily move back and forth between languages with no perceptible differences in the game itself other than the menus and dialogue. This variation produces a game that seems to be simultaneously made for multiple linguistic populations, but it also disables the possibility of extensive localization of content, graphics, sound or music. Because the switch is flipped when the game is already loaded and being played, different gameplay, graphical and audio assets are not loaded, which forces these features to remain uniform for all users. The only thing that changes is whether Japanese or English is written on the screen.

Both types of multilingual switch render visible multiple languages at the same or less cost than highly localized monolingual games and they also allow the player to switch easily between languages, thereby actively playing with their own linguistic and cultural identity. In the case of diasporic players, the ability to choose their language is of crucial importance. However, there are two issues associated with multilingual versions: censorship and cost.

12 *Chaos Rings* is an interesting example as Square-Enix's previous games utilize the iPhone OS's language to determine the game's language (the second monolingual option), and it's most recent game, *Seiken Densetsu 2* (2010), uses the title screen switch method (the first multilingual variation).

Certain locations require specific features such as imagery and violence to be censored.[13] A game with Nazi representations might never be released in Germany if it were to have a multilingual switch that did not enable a more extensive form of localization. Even with extensive localization, games can have problems with censorship issues: the version of Raven and id Software's remake *Wolfenstein* (2009) released in Germany was recalled due to one remaining swastika missed by the developers, as the law in Germany necessitates the removal of all Nazi imagery within non-historical entertainment products (Mastrapa, 2009). Because the German versions of games need all Nazi representations to be removed, the existence of these in other localizations on the same disk (as either audio dialogue or graphical assets) would hinder a game's release, making a multilingual language implementation difficult or impossible. The result of stringent requirements such as censorship within a target locale sometimes necessitates a monolingual language implementation containing the single language on the disk, as other linguistic environments have not been appropriately censored to conform to the rules of the target location. Unfortunately, as discussed earlier, monolingual localization is problematic in terms of alternate groups and accessibility, as it produces a singular environment that does not exhibit or allow any difference.

The second problem with multilingual localization involves the economic costs and benefits. To smaller companies wanting to break into the international market, multilingual variations might be the better option because of the costs of development, but to larger companies the monolingual, highly localized option is considered to result in better sales. Even though a monolingual localization might cost more, the belief is that a single game tailored to an audience sells more. This cost-benefit difference helps to support a hierarchy between different styles of language implementation due to the perceived level of immersion, where the monolingual localization is seen as the most immersive (and therefore best) option. However, as monolingual games fail to be fully accessible to alternate groups such as diasporic populaces, the determination of "best" must be ques-

13 An issue that is skipped here is the requirement of certain languages. The politics of national and minor languages are important and visible in decisions of certain companies to release games particular languages on the disk. The decision to have Spanish included in a game to be released in the United States is related to geopolitics, as is the decision not to include Spanish (or Chinese). Related to this is the mandate to include French in games released within Qúebec (Chung, 2009).

tioned and other options considered. Multilingual games remove Edwards' prop of similarity (2001) enabling the constructive difference needed for diasporic groups, but are difficult to produce as they face political and cultural issues such as censorship or simply difficulty understanding what is appropriate to a particular context. They also require developer effort instead of capitalizing on user production which has become so prevalent with digital texts of the 21st century (Jenkins, 2006).

Omnilingual Language Implementation

The final category of language implementation is 'omnilingual'. Like multilingual implementation, omnilingual implementation can use multiple languages, but sacrifices *full localization* strategies including graphical and gameplay alterations as well as censorship and culturalization (Edwards, 2006).[14] While multilingual implementation includes several languages to fit a context (the region of North America generally has French, English and Spanish), omnilingual implementation tries to encourage a type of 'drag and drop' methodology that could fit almost any language and could be utilized by many. Examples of this are the Polish game company CD Projekt RED STUDIO's *Witcher* (2008), and independent game company Basilisk Games' plans for *Eschalon: Book II* (2010). It must be noted that both games are primarily available for PC/Mac systems, which enable certain possibilities disabled by the more restrictive Nintendo, Sony and Microsoft consoles. However, there is no material reason why this type of implementation cannot exist on console systems, particularly with the increased reliance on downloadable content (DLC) of the current game industry strategies.

14 Censorship is the government enforced determination of what is acceptable within a game, and culturalization is the developer and localizer-led alteration. An example of censorship is the previously discussed *Wolfenstein* (2009). An example of culturalization is the "No Russian" level in *Call of Duty: Modern Warfare 2* (2009). This controversial level involves a massacre of civilians in an airport (Famitsul 2009; Warmoth. 2009). The localized variations of the level include a) making the player decide whether or not to shoot civilians, but making it impossible to save the civilians (United States etc), b) making it impossible to shoot the civilians (Japan, Germany), and c) removing the level altogether (Russia). The level is an example of a developer-led effort at culturalization where they altered the gameplay of the level based on assumptions as to what the different local audiences would deem acceptable.

The expanded edition of *Witcher* incorporates many different subtitle and dialogue tracks allowing the user to mix and match between them. Because CD Projekt RED STUDIO is a Polish developer surrounded geographically by a host of European countries with a range of languages, but also partially because of the unexpected popularity of the game, the expanded edition of *Witcher* attempts to extend the game's reach as far as possible through the implementation of numerous subtitle and dialogue tracks. However, the distribution of an edition of the game that enables many different linguistic options seems to mirror CD Projekt's history as a game localization company for the Eastern European market.[15] Similar to the drive to release mainstream English-language titles into the Eastern European market, the expanded edition of *Witcher* is intended to be a far-reaching game. However, the implementation is unique in its affordances: English voice dubbing and subtitles, Polish voice dubbing and dialogue, Polish voice dubbing and English subtitles, Italian voice dubbing and Spanish subtitles, or any number of other possibilities. While the game's content does not change, that is to say there are no graphical or gameplay changes between versions released to different locales, it nevertheless has an unexpectedly broad global spread due to the unique language implementation it employs.[16] The popularity of the game and the success of its language implementation strategy should encourage discussion as to whether monolingual language implementation is actually as good as it is generally considered to be within the industry and associated discourse. Perhaps less fluidity has its benefits?

A second type of omnilingual language implementation can be seen in the second title of Basilisk Games' indie-RPG *Eschalon* series. According to the developer forums, there are plans to facilitate user-created language packs to enable different languages as downloadable content (Basilisk Wrangler, 2010). The developers plan to incorporate the utilization of fan production by allowing users to download and translate the game script. The new, fan-produced language packs could then be downloaded and integrated by other players. As with the multilingual edition, this pack would be switchable within the game. Although the language packs are

15 Much of CD Projekt's history is documented on the company's website: <http://www.cdprojekt.com/company.asp?id=g_history>.

16 Of course, this implementation is not unique, as it replicates the audio and subtitle options of DVDs, but it is rare within the video game medium.

not integrated as yet, their implementation would likely only require what is currently considered by localization standards to be good practice, namely the organized integration of different languages into a nested file structure (Chandler, 2005 and 2009). As long as the game has been programmed with such a structure, meaning that the game can load the appropriate language files at the appropriate time, the implementation should not be difficult.

The main difficulty in producing an omnilingual language implementation is not with the technical elements, as the only technical necessity involves planning during the production cycle in order to have a nested file structure that allows the program to grab certain audio, visual, and gameplay assets depending on the language utilized. Rather, the main difficulty lies in encouraging developers, producers and localizers to make games in this manner. To larger companies, the monolingual language implementation has 'better' discursive and monetary support. To many smaller companies that do not initially plan to produce alternate localizations, the creation of a particular file structure might be seen as a meaningless additional requirement in the coding process. However, one can consider Basilisk Games – an independent producer with one permanent staff member and a large number of fans who would gladly do the work – as an example of a small company that has benefited from taking that additional step. Because the game was coded with a nested file structure (involving a small investment of skill and time) Basilisk Games is now able to obtain a much larger global spread through utilizing fan production – something that is possible with any game.

Furthermore, the size of the company should not impede omnilingual language implementation. The process simply requires the utilization of the affordances provided by new media production (including distributed production). One way to understand such new media production is through what Howe calls 'crowdsourcing' (2006). An example of crowdsourced localization is when Facebook employed individual users to translate phrases in a game-like environment when it localized its online social networking site to other languages and locales. While there are numerous critiques of some of crowdsourcing's more corporatist instantiations (such as Amazon's Mechanical Turk (Aytes, 2011)), there have been fewer ethical qualms raised concerning the independent translation method of FLOSS (Free Libre Open Source Software) manuals. This process capitalizes on what Benkler (2006) calls "peer production" whereby people do things for different reasons (pride, cultural interaction, economic advancement etc). Here, crowdsourcing seems to avoid the standard critiques relating to the payment of low wages and deskilling of workers. Rather, it is about doing something for the love of the product, just as Jenkins (1992) noted for fan production two decades ago. The utilization of non-industry means can enable alternate forms of production, and this production serves alternate audiences.[17] One such audience (and source of labor) is that of the various diasporas.

6. Moving Towards Diasporic Games

Through current localization practices, games force a player into a particular position by reifying as natural the borders and locales produced through localization. This holds true with any player in any locale, be it the Mandarin speaking Chinese 'locale', the English speaking North American 'locale', the German/French/Spanish speaking European 'locale', or any number of other determined 'locales' that do not fully satisfy the subjectivity of a player, and do not even come close to representing the subjectivity of players in diaspora. In the case of a diasporic Chinese person in America (or many other geographic locations of the Chinese diasporas), his or her determined subject position is away from the Sinophone due to the limited languages contained in a North American localization. In the case of a Cantonese speaking Chinese mainlander, it is into the position of a Mandarin speaking, imperial subject that he or she falls, due to limitations of traditional and new scripts and of dubbing tendencies that wipe out difference under a homogenous topolect. These forced positions are

17 Naturally there are serious issues regarding labor here on two fronts. The first is in the labor of the fans creating translations. This is alleviated through micro-payments for the additional localization packages. They must receive some amount of compensation for their labor, as this situation is dangerously close to exploitation. The second issue is related to the de-skilling of professional translators and localizers due to the possible disappearance of their work to the fan translators. This is an issue, but micro-payments and the necessity of companies to pay localizers for the primary localizations should somewhat alleviate this possible de-skilling. It is, however, a matter that demands attention.

very difficult to escape from, as software developers and publishers usually police their copyright by blocking all manipulation of the final coded application including alternative, independent and fan translations.[18] Such alternate translations create new playing positions but the current industry system blocks their release, if not their existence, thereby maintaining singular positions.[19] In order to create a diasporic game with the multiplicity of viewpoints necessary to interact with the Sinophone, the author proposes the omnilingual language implementation method described above, but organized to utilize fan production.

A Sinophonic game would be easier to produce than a fully omnilingually diasporic title, as the written language would actually stay the same between versions making the effort to bring out the diasporic nature a primarily audio based one. In contrast, the true omnilingual game would need to have all semiotic registers available for alteration, including those relating to graphics and gameplay (which are much more difficult to change). These extended registers are beyond the responsibility of most localization studios and within the purvey of the developers and publishers even though they reinforce the idea of localization. While the creation of the Sinophonic game is no mean task (especially with the sheer mass of dialogue in the average game) it is a good first step towards creating other diasporic titles.

The technological requirements for creating a Sinophonic game are a nested file structure and the permitting of changes to audio files. One example where both of these are possible is in Bioware's *Baldur's Gate 2* (2000), which had specifically labeled main character dialogue assets, and gave the user instructions as to how to replace the standard files with ones

18 KWhazit's *Chrono Trigger* retranslation is one example. Of particular note is the 'cease and desist' notice from the game's publisher Square-Enix on the retranslation's hosting website. See: <http://www.chronocompendium.com/Term/Retranslation.html>. Some other fan translations that have been made are *Mother 3* <http://mother3.fobby.net/> and *Seiken Densetsu* 3: <http://www.neillcorlett.com/sd3/>. One possible reason that the latter two translations have not been aggressively halted by the publishers is that they have never been officially localized or released into the English-language markets.

19 Blocking alternate positions and visions is particularly visible with the console system where a centralized company (Nintendo, Sony, and Microsoft) supervises the games released and played on the system in order to control its corporate image. In contrast, and represented by the author's sample set that includes a predominant number of PC games, the PC market has less stringent requirements and thus has many more interesting things are happening.

that he or she had created, thus replacing the regular main character's pre-programmed voice with the player's own. There are also numerous downloadable fan-created packs online that allow the user to avoid doing the work of recording herself.[20] However, unlike in *Baldur's Gate 2*, the Sinophonic game would allow the alteration of the entire register of dialogue files including the NPC dialogue, sound effects, and background music, instead of simply the main character's battle cries. The Sinophonic game might come with standard Mandarin as the officially localized language, but it would allow users to translate the language as desired in order to create their own game world, and these user localizations would be downloadable through the developer's website.[21] While recreating dialogue files with a particular dialect would take a large amount of effort, it would be developer/publisher sanctioned but ultimately fan made, and would allow multiplicity and interaction with difference. As the various aforementioned examples show, this enthusiastic labor base is readily available. It is this active involvement of changing the modular[22] assets of a game that can bring out an idea of diaspora.

7. Conclusions

This paper has sought to elaborate on a basic problem with video game localization: that it fails to make games accessible to alternate populaces such as communities in diaspora. Because localization reifies a particular link between location, language and player subjectivity, any differentia-

20 For examples of fan efforts to adapt the *Baldur's Gate 2* voices see: <http://www.sorcerers.net/Games/BG2/index_soundsets.php>. For an effort to completely re-record the dialogue of the game see: <http://voiceacting.proboards.com/index.cgi?board=general&action=display&thread=19681>.

21 While here a PC environment is envisaged, a console implementation should be equally possible due to the existence of downloadable content and in-app purchases within many console environments.

22 Here, 'modular' is one of Lev Manovich's (1999) principles of new media. Modularity, along with numerical representation, automation, variability and transcoding are the principles of new media. The changeable assets are modular elements of the game, so changing them is a natural part of the medium and method for bringing out an idea of diaspora.

tion causes accessibility problems. The author has used as an example the members of the Chinese Diaspora who practice their subjectivity through a non-standard combination of location and language, but the issue stretches to any group that does not combine location, language and player subjectivity in the assumed manner that has become standard within localization. As a means of solving this problem of access, using the "omnilingual language implementation" visible in some non-standard video game localizations has been offered as a methodology for better inclusion.

The proposed diasporic game would both connect and differentiate people in an accessible way. While it might be seen as a utopian thought experiment as it goes against dominant industry practice, it is neither impossible nor unreasonable. The availability of fan production and the way this can fit into localization practice shows that the diasporic game is a real possibility. Furthermore, the diasporic game can serve as a model for what might be possible with the full utilization of a medium to represent, and be represented by, an alternative group: it can show what is next, and what is needed, in localization practice.

References

Appadurai, Arjun. 1996. *Modernity at Large: Cultural Dimensions of Globalization, Public Worlds*. Minneapolis: University of Minnesota Press.

Ashcraft, Brian. 2010. 'Why Are Consoles Banned in China?'. Online at: <http://kotaku.com/5587577/why-are-consoles-banned-in-china> (consulted 07.12.2010).

Aytes, Ayhan. 2011. 'Cognitive Labor, Crowdsourcing, and Cultural History of the Mechanization of the Mind' in *Leonardo Electronic Almanac* Vol. 17, No. 1: 118–127.

Basilisk Games. 2010. *Eschalon: Book II*. Basilisk Games. BasiliskWrangler. 2010. 'Status of version 1.04, and the future…'. Online at: <http://basiliskgames.com/forums/viewtopic.php?f=12&t=3818> (consulted 05.07.2013).

Benkler, Yochai. 2006. *The Wealth of Networks: How Social Production Transforms Markets and Freedom*. New Haven: Yale University Press.

Bioware. 2000. *Baldur's Gate II: Shadows of Amn*. Black Isle Studios; Interplay Entertainment.

Blizzard Entertainment. 2004. *World of Warcraft*. Blizzard Entertainment.

Blizzard North. 2000. *Diablo II*. Blizzard Entertainment.

Capybara Games. 2009. *Might and Magic: Clash of Heroes*. Ubisoft.

CD Projekt RED STUDIO. 2008. *The Witcher: Enhanced Edition*. Atari.

Chandler, Heather Maxwell. 2005. *The Game Localization Handbook*. Hingham, Mass.: Charles River Media

Chandler, Heather Maxwell. 2009. *The Game Production Handbook* (2nd ed.). Hingham, Mass.: Infinity Science Press.

Chung, Andrew. 2009. 'New Quebec Law Turns Lara Croft Into Francophone.' Online at: <http://www.thestar.com/article/611472> (consulted 05.07.2013)

Clifford, James. 1997. *Routes: Travel and Translation in the Late Twentieth Century*. Cambridge: Harvard University Press.

Díaz, Junot. 2007. *The Brief Wondrous Life of Oscar Wao*. New York: Riverhead Books.

Edwards, Brent Hayes. 2001. 'The Uses of *Diaspora*' in *Social Text* 66, Vol. 19, 1. Duke University Press.

Edwards, Kate. 2006. 'Fun Vs. Offensive: Balancing the "Cultural Edge" of Content for Global Games.' Paper presented at *Game Developer's Conference: What's Next*. San Jose.

Entertainment Software Association.2011. 'US Copyright Industries Highlight Intellectual Property Rights Deficiencies in Foreign Markets.' Online at: <http://www.theesa.com/newsroom/release_detail.asp?releaseID=136> (consulted 05.07.2013).

Famitsu. 2009. 'Call of Duty: Modern Warfare 2 ha nihongo onsei, jimaku no mi.' Online at: <http:.//www.famitsu.com/game/news/1230281_1124.html> (consulted 05.07.2013).

Flanagan, Mary. 2007. 'Locating Play and Politics: Real World Games & Activism'. Paper presented at the Digital Arts and Culture, Perth, Australia.

Fresh, DJ Fob. 2010. 'A Yakuza 3 Guide to Edits and Cuts.' Online at: <www.segashiro.com/2010/03/11/a-yakuza-3-guide-to-edits-and-cuts/> (consulted 05.07.2013).

Galloway, Alexander R. 2006. *Gaming: Essays on Algorithmic Culture*. Minneapolis: University of Minnesota Press.

Gilroy, Paul. 1993. *The Black Atlantic: Modernity and Double Consciousness*. Cambridge, Mass.: Harvard University Press.

Infinity Ward. 2009. *Call of Duty: Modern Warfare 2*. Activision.

Insomniac Games. 2002. *Ratchet and Clank*. SCEI.

Jenkins, Henry. 2006. *Convergence Culture: Where Old and New Media Collide*. New York: New York University Press.

Jenkins, Henry. 1992. *Textual Poachers: Television Fans & Participatory Culture, Studies in Culture and Communication*. New York: Routledge.

Kilinc, Oktay, et al. 2006. 'How to Change the Game Language in Diablo II Lord of Destruction'. Online at: <http://www.wikihow.com/Change-the-Game-Language-in-Diablo-II-Lord-of-Destruction> (consulted 13.07.2010).

KWhazit. 2007. Chrono Trigger Retranslation. Online at: <http://www.chronocompendium.com/Term/Retranslation.html> (consulted 09.12.2010).

Lee, Ang. 2000. *Crouching Tiger, Hidden Dragon*. EDKO Film and Sony Pictures Classics.

Liao, Ping-hui. (forthcoming). 'Introduction: Here or There, at Home Everywhere?' to appear in Liao (ed.) *Global Diasporas and International Critical Studies*.

Lin, Holin and Chuen-Tsai Sun. 2011. 'A Chinese Cyber Diaspora: Contact and Identity Negotiation on Taiwanese WoW Servers.' Paper presented at *Think Design Play: 5th Annual DiGRA conference*, Hilversum, Netherlands.

Manovich, Lev. 1999. *The Language of New Media*. Cambridge: MIT Press.
Mastrapa, Gus. 2009. 'Wolfenstein Recalled in Germany, Inglourious Basterds Still Fine'. Online at: <http://www.wired.com/gamelife/2009/09/wolfenstein-recalled-in-germany/> (consulted 05.07.2013).
Media Vision. 2010. *Chaos Rings*. Square Enix.
Mirzoeff, Nicholas. 1999. 'Introduction: The Multiple Viewpoint: Diasporic Visual Culture' in Nicholas Mirzoeff (ed.) *Diaspora and Visual Culture: Representing Africans and Jews*. London and New York: Routledge.
Raven, id Software. 2009. *Wolfenstein*. Activision.
Shih, Shu-mei. 2007. *Visuality and Identity: Sinophone Articulations across the Pacific*. Berkeley: University of California Press.
Shih, Shu-mei, Chien-hsin Tsai and Brian Bernards (eds) (forthcoming). *Sinophone Studies: A Critical Reader*. New York: Columbia University Press.
Square Enix Co., LTD. 2010. *Seiken Densetsu 2*. Square Enix Co., LTD.
The Entertainment Software Association. 2011. 'US Copyright Industries Highlight Intellectual Property Rights Deficiencies in Foreign Markets.' Online at: <http://www.theesa.com/newsroom/release_detail.asp?releaseID=136> (consulted 03.10.2011).
Tölöyan, Khachig. 1996. 'Rethinking Diaspora(s): Stateless Power in the Transnational Moment' in *Diaspora, 5*(1).
Trading Markets.com. 2010. 'Sony Winks at PSP Rampant Piracy in China.' Online at: <http://www.tradingmarkets.com/news/stock-alert/sne_sony-winks-at-psp-rampancy-in-china-1052049.html> (consulted 21.03.2011).
Tu, Wei-ming. 1991. 'Cultural China: The Periphery as the Center.' in *Daedalus* 120, 2: 1–32.
Urry, John. 2007. *Mobilities*. Cambridge: Polity.
Wang, Gungwu. 1999. 'Chineseness the Dilemmas of Place and Practice' in Hamilton, Gary (ed.) *Cosmopolitan Capitalists: Hong Kong and the Chinese Diaspora at the End of the 20th Century*. University of Washington Press.
Warmoth, Brian. 2009. 'No Russian' Scene Censored for 'Call of Duty: Modern Warfare 2' In Japan.'Online at: <http://multiplayerblog.mtv.com/2009/12/09/no-russian-scene-censored-for-call-of-duty-modern-warfare-2-in-japan/> (consulted 05.07.2013).

Oliver Carreira & Eugenia Arrés

Video Game Localization Training on Offer in Spanish Universities at an Undergraduate Level

The video game localization market is currently rapidly expanding, meaning that an increasing number of qualified professionals is required. Specific training is needed to meet the market standards, but it seems that the number of undergraduate level (licenciatura *or* grado) *localization courses on offer at Spanish universities is relatively limited when considered in absolute terms.*

In this paper, we aim to analyze the apparent gap between the academic and professional environments with regards to the field of video game localization. Firstly we focus on some of the prerequisites for working in this field. Following this assessment, we analyze the training on offer. We then proceed to interview a number of localization professionals in order to garner their opinion on the training currently offered.

Finally, with the information gathered, we propose several strategies for introducing game localization into undergraduate curricula, in the hope that these strategies may contribute to bridging the gap between the Spanish educational framework and the employment market.

1. Introduction

The importance of both the video game and video game localization markets has increased in recent years. We believe that both these sectors offer a large number of professional opportunities for translation and localization professionals. Despite this, it seems that the specific training offered by Spanish universities in the Translation and Interpreting curricula at an undergraduate level does not reflect this situation. Whilst we found other areas of specialization such as medical, legal and technical translation were covered by the curricula, this was not the case for localization (and specifically for video game localization).

The objective in this paper is to investigate this situation and to determine the training currently provided by institutions of tertiary education. Firstly, we give an insight into the localization and video game markets in order to highlight the relevance of these fields in the present socio-economic context. Next, we consider the situation of undergraduate degrees related to translation studies which are offered by Spanish universities, and the specific features of the Bologna Process in Spain. Following these two introductory sections, we provide the results of a survey of video game localizers, in which we attempted to ascertain whether this group of professionals received specific training at university for their job.

Considering the aforementioned data, we provide some conclusions related to our research, ending with several proposals to promote a change in the current curricula models.

We are aware that most of the specialized training in this field is offered at postgraduate level. Despite this, we have decided to analyze the situation at undergraduate level for two reasons. Firstly because the situation at postgraduate level has already been investigated (Vela, 2011), and secondly because we firmly believe that undergraduate programmes *must* offer a panorama of all the different specializations available in the field of translation and, considering the importance of the video game localization market, video game localization should be included as one of these.

2. What is Video Game Localization?

For the purpose of this study, we will consider the traditional idea of localization as being the process of adapting a product or service to a particular language and culture. However, certain nuances are always necessary to distinguish between translation and localization. Importantly, the localization process takes into account the following aspects, which are not always relevant in standard intralingua translation (Lommel, 2007: 12–15):

- *Linguistic issues*: the linguistic adaptation of the product which is going to be sold to individuals (i.e. the product itself, marketing and collateral product materials, web pages, support materials, dubbing and the adaptation of speech-based audio components)

- *Physical issues*: the physical adaptation of the product to the specific market (i. e. voltage, computer keyboard layouts)
- *Commercial and cultural issues*: the adaptation to local business and culture of product design and localization (i. e. currencies and accounting conventions, address and telephone number formats, colour and graphics)
- *Technical issues*: special attention and planning at the engineering stage (i. e. support for East Asian languages that require thousands of characters, support for languages that are written from right to left such as Arabic and Hebrew)

These special requirements should always be met in order that the product be acceptable in the local market – they allow companies to save time and costs, reducing the process of quality assessment required.

3. The Relevance of Localization in the Current Undergraduate Spanish Curricula

This issue was raised during the course of the implementation of the Bologna Process[1], when Spanish Translation schools gathered to plan their new curricula. At that moment (2004), as stated by the common draft named *Libro Blanco*, the development of both new technologies applied to translation and computer-assisted tools was uncertain[2].

1 The aim of the Bologna Process is to create a European Higher Education Area that promotes mobility, attracts students and staff from Europe as well as from other parts of the world by facilitating greater comparability and compatibility between the diverse higher education systems and institutions across Europe and by enhancing their quality. For more information on this subject, see Kettunen and Kantola (2006).

2 The draft stated the following regarding the evolution of technologies and its application to curricula: "A la espera de desarrollos todavía imprevisibles en el ámbito de las nuevas tecnologías y de las herramientas de traducción asistida por ordenador, todas y cada una de las competencias consideradas imprescindibles se hallan recogidas en este estudio, avaladas por el consenso al que han llegado las Facultades y los Departamentos de Traducción e Interpretación de España y las diversas asociaciones profesionales, tanto nacionales como internacionales." (Libro Blanco, 2004)

With the Bologna Process, the subjects within the curricula of Translation schools were divided into several categories. For the purpose of this research, we will assess two of these:

1) Basic training (compulsory for all Translation schools): subjects such as "Informática Aplicada a la Traducción" (IT applied to translation) and "Nuevas Tecnologías Aplicadas a la Traducción" (New technologies applied to translation).
2) Specialized training (comprising a number of elective subjects depending on the university): subjects such as "Localización" (Localization) and "Traducción de Software y Páginas Web" (Web and software translation). The aforementioned uncertainty regarding the development of translation technologies affects these new curricula, as not all undergraduate courses in translation (known in the new Spanish educational system as *Grados*) include subjects related to specialized training in localization.

However, some of the specific competences related to the localizer profile were included in the draft written by the Translation schools in the form of a general list of essential competences:

1. Foreign language proficiency.
2. Knowledge of foreign cultures and civilizations.
3. Spanish language proficiency (written and oral).
4. Proficiency in the techniques and terminology of specialized translation.
5. IT tools proficiency.
6. Proficiency in assisted translation/localization techniques.
7. Competence in searching for information/documentation.
8. Knowledge of professional and economic features.
9. The ability to work effectively as part of a team.
10. The ability to design and manage projects.
11. A broad cultural background.

As we can see, localization is specifically mentioned as one of these key competences. Furthermore, when students were asked about the specific competences they considered important for their professional activity in a survey that formed part of the research carried out for the *Libro Blanco*, two of the highest ranked competences were those of computer-assisted tools and expert localization knowledge[3]. In this sense, it seems that there is a gap between what is offered by universities and what students consider to be most important for their training.

One of the main aims of the research reflected in the *Libro Blanco* was to take into account the need to meet the demands of the translation market for new graduates. This document also gathers together some recommendations regarding the market from professional associations such as AIIC, ASETRAD and TRIAC among others (ANECA, 2004: 102). These associations consider the following measures to be essential:

- Specialization in translation
- More professional translators teaching at Translation schools
- Reinforcing professional internships in academia
- The incorporation into translation training of practical aspects from professional practice
- The reinforcing of certain training areas (e.g. new technologies applied to translation)

The recommendations made by the Spanish Translation schools and professional associations are based on market research, their aim being to detect and/or confirm that the localization market is a growing industry.

Ever since its inception, and probably motivated by the fact that many companies needed translation services to adapt their local products to foreign markets, the localization industry has been constantly developing and evolving, and a greater level of growth is expected in the coming years. Therefore, if we consider the localization market to be an interesting niche for translation graduates, it is obvious that the community of linguistic professionals can benefit from this increase in activity.

In order to illustrate this, below is shown the projected language services revenues ranking of the 30 most important language service companies, conducted by Common Sense Advisory in 2009.

3 The survey shows the following results: "En cuanto a las competencias específicas, destaca en los traductores el dominio la traducción asistida y la localización, la capacidad de trabajar en equipo, la de organización del trabajo y gestión de proyectos, el razonamiento crítico y, naturalmente, la excelencia en todas las lenguas de trabajo, incluida la propia."

Region	Market Share	2009 US$ M	2010 US$ M	2011 US$ M	2012 US$ M	2013 US$ M
Europe	43%	6,468	7,331	8,409	9,703	10,781
U.S.	40%	6,074	6,884	7,896	9,111	10,123
Asia	12%	1,735	1,965	2,255	2,601	2,891
ROW	5%	722	818	939	1,083	1,203
Growth Totals	100%	15,000	17,000	19,500	22,500	25,000

Table 1: Projected language service revenues for 2009–2013 in millions of dollars (US). Source: Common Sense Advisory, Inc.

In this table, two important facts can be highlighted:

1) The language service industry (which employs a great deal of translation graduates as translators, proofreaders, terminologists, project managers, editors, etc.) is expanding and a significant level of growth is predicted in coming years.
2) The predictions for growth are slightly lower than those published before the global economic crisis, but a compound annual growth rate of 10.76% is still expected for the period 2009–2013.

According to the research carried out by Common Sense Advisory into the top 30 language service providers worldwide, most of these companies, (and especially those ranked in the top ten) largely devote themselves to providing localization services. In addition to the older companies represented which are recycling their services to the new technologies market, some of the newer companies appearing in the table are also starting to offer localization services.

Regarding the different tasks and positions held within these companies by translation graduates and the competences previously mentioned, we refer to Figure 1, which explains the importance of localization as being complementary to the additional competences necessary for translators.

Localization is a complex process involving a wide range of activities which can be itemized in different competences and in several processes. As illustrated by Wordbank Ltd. in the following graph reflecting two years of accumulated data from localization projects, "translation is only a part of the localization process".

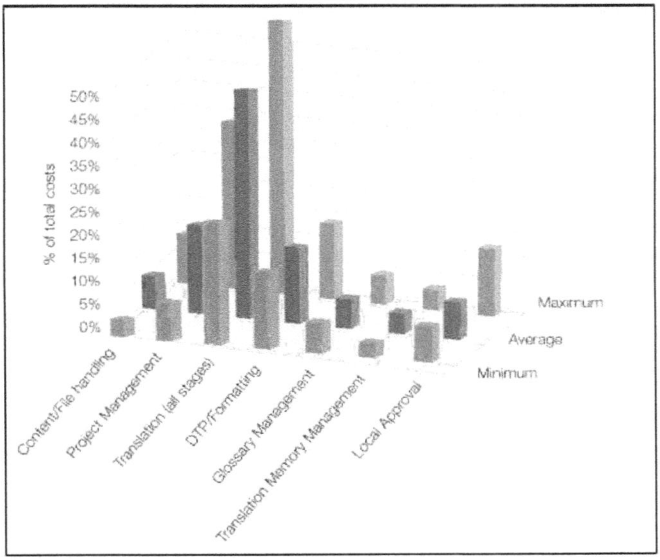

Figure 1: The total cost of localization. Source: Wordbank Ltd.

Besides the costs devoted to translation (which only account for 40% of the total expenditure), we can also observe the other profiles of content/file handling, project management, DTP/formatting, glossary management, translation memory management and local approval. Every task mentioned could be performed by a translation graduate, so it would seem that the localization process would be a good way to incorporate professional competences into Translation schools through academic training.

4. The Situation of the Video Game Market

The main characteristics of the video game market are its complexity and segmentation. Company mergers or reorganization of different business divisions is quite usual and takes place according to the economic and strategic objectives of every company. This makes research into this market difficult, as it is necessary to consider a complex flux of information and several variables. For example, all the economic figures must be converted into a single currency. Some of the data are in Yen, others in Dollars and others in Euros. Furthermore, the conversion rate used for comparison must be the one from the day on which the information was published. Another variable to consider is inflation. In order to obtain comparable values, the figures for each date must be converted to an inflation-equivalent value.

The analysis performed concerning the video game market is based on the comparison and study of the net income values of the following companies for the period 2003–2010: Activision Blizzard, Electronic Arts, Konami, Microsoft, Nintendo, Sega Sammy, Sony, Square Enix and Ubisoft. We have worked with the annual reports published by these companies, as well as the available tax information. The selection of these particular companies (and not others) is based in two criteria: their economic weight in the market and the availability of information for the period considered. Though we are aware that our analysis does not cover 100 % of the video games market, we believe that the results allow us to accurately depict the overall market situation.

We can clearly distinguish between two segments within the video game market. On the one hand, there are those companies that both manufacture consoles and develop video games: Nintendo, Sony and Microsoft. These companies have experienced a positive and constant growth rate throughout the period examined, with net income values in 2010 ranging between $2,000 and $6,000 million (US).

In the second segment are those companies that only develop and distribute video game software for the platforms developed by the companies in the first segment. Their net income values are much lower, ranging between $25 and $400 million (US). Within this second segment we can also distinguish two subsegments. Subsegment A includes those companies that experience a positive and constant growth rate during the period, ranging between $25 and $200 million (Activision Blizzard, Konami, Square Enix and Ubisoft). Subsegment B includes Sega Sammy and Electronic Arts. These companies suffer a greater level of instability, with prominent changes between scenarios of high profit (almost $600 million) and significant losses (around -$1,000 million).

If we analyze the interperiod average growth, all the values are positive. Sony is the best positioned company in the platform development segment, with a net income value of $10,956 million, while Activision holds the number one position in the software development segment, with a net income value of $144.63 million.

The global financial crisis does not seem to have affected the video game market. The platform development segment did not experience any drop at all. All the companies in this segment experienced spectacular growth (excluding Nintendo, which did grow during the period but at a slower pace). On the other hand, the software development segment shows some signs of instability. Sega Sammy and Electronic Arts experienced significant losses post-2007. However, both companies are now back to having positive results, the Sega Sammy after 2008 and the Electronic Arts after 2009.

5. Educational Resources Related to Video Game Localization in the New Spanish Curricula

Before describing the educational resources in the Spanish curricula, it seems pertinent to clarify the transitional situation we are currently experiencing within the Spanish university system as a result of the implementation of the Bologna Process. This has made researching the different resources used in the new undergraduate courses or *Grados* a little difficult. In the curricula of some of the *Grados en Traducción e Interpretación*, the planning of the new courses that will be offered in the last year of the degree (mainly the 3rd or 4th years) have not yet been developed, which in some cases prevented us from acquiring all the necessary information about the courses.

The requirements of the new curricula established with the implementation of the *Grados*, and the possibility of reviewing the previous curriculum offered to the Spanish translation departments through the Bologna Process, have demanded a course selection and reorganization in the *Grados* of each Translation school. Based on this need for adaptation, less traditional and more technologically-oriented courses have been added at some of these training centres.

With the aim of establishing which educational proposals relating to video game localization are implemented in Spanish translation schools and how useful these are for the professional activity of their graduates, we have based our research on a selection of the Spanish universities which offer the *Grados* in Translation and Interpreting. The collection of relevant data has been carried out for all the official *Grados* of every university school or department offering this degree. The exception to this is the *Dobles Grados*[4], as we have observed that in all cases the courses which are supplementary in a combined second degree are not specifically related to new technologies or localization, and they therefore do not contribute any new data to this paper.

After studying the curricula, we have extracted those courses we considered more likely to include some content related to video game localization, as they were directly related to localization or new technologies applied to translation. For the purpose of this research, we decided to select only those courses which could include some video game localization training as "direct training", considered as a full theoretical and practical training with the aim of providing students with the knowledge and competences related to video game localization. We therefore discarded all those courses that could incorporate video game localization foundations as a complementary training for other purposes such as, for example, using a dialogue text from a role playing game (RPG) to study some syntactic aspects of the source language (i.e. in a grammar subject).

In the following table (Table 2), we outline the main details of each of the schools analysed in our study and include the name of the course(s) selected for the purpose of this research. For this, we have selected the courses directly related to translation technology and computer aided translation (CAT) tools, new technologies, and multimedia translation and localization.

4 The degree called "Doble Grado" is in fact a combination of two independent degrees from two different disciplines (i.e., Degree in Translation + Degree in Humanities).

University	*Public / Private*	*Degree*	*Course/s selected*
Universidad de Alicante	Public	Grado en Traducción e Interpretación (Four-year degree in Translation and Interpreting)	1. Tecnologías de la Traducción (Technologies in Translation)
Universidad del País Vasco/Euskal Herriko Unibertsitatea	Public	Grado en Traducción e Interpretación (Four-year degree in Translation and Interpreting)	1. Informática Aplicada a la Traducción (IT Applied to Translation)
Universitat Pompeu Fabra	Public	Grado en Traducción e Interpretación (Four-year degree in Translation and Interpreting)	1. Informática y Documentación (IT and Documentation) 2. Tecnologías de la Traducción (Translation Technology) 3. Localización (Localization)
Universitat de Vic	Private	Grado en Traducción e Interpretación (Four-year degree in Translation and Interpreting)	1. Informática I (IT I) 2. Informática II (IT II)
Universitat Autònoma de Barcelona	Public	Grado en Traducción e Interpretación (Four-year degree in Translation and Interpreting)	1. Introducció a les Tecnologies de la Traducció i de la Interpretació (Introduction to Technologies in Translation and Interpreting) 2. Tecnologies de la Traducció i la Interpretació (Technologies in Translation and Interpreting Studies)
Universidad de Córdoba	Public	Grado en Traducción e Interpretación (Four-year degree in Translation and Interpreting)	1. Herramientas Informáticas Profesionales (Professional IT Tools) 2. Herramientas Profesionales para la Traducción (Professional Tools for Translation) 3. Informática Aplicada a la Traducción (IT Applied to Translation) 4. Traducción Multimedia (Multimedia Translation)
Universitat Jaume I	Public	Grado en Traducción e Interpretación (Four-year degree in Translation and Interpreting)	1. Noves Tecnologies per a les Llengües i les Humanitats (New Technologies for Languages and Humanities) 2. Tecnologies de la Traducció (Technologies in Translation)
Universidad de Granada	Public	Grado en Traducción e Interpretación (Four-year degree in Translation and Interpreting)	1. Informática (IT) 2. Traducción Multimedia (Multimedia Translation)
Universidad Alfonso X El Sabio	Private	Grado en Traducción e Interpretación (Four-year degree in Translation and Interpreting)	1. Informática Aplicada a la Traducción (IT Applied to Translation) 2. Traducción y Nuevas tecnologías (Translation and New Technologies) 3. Traducción Científico-técnica y Localización (Technical-Scientific Translation and Localization)

Universidad Autónoma de Madrid	Public	Grado en Traducción e Interpretación (Four-year degree in Translation and Interpreting)	1. Informática Aplicada a la Traducción (IT Applied to Translation) 2. Informática Avanzada para Traductores (Advanced IT for Translators) 3. Localización de Software y Programación Web (Software Localization and Web Programming)
Universidad Europea de Madrid	Private	Grado en Traducción y Comunicación Intercultural (Four-year degree in Translation and Intercultural Studies)	1. Traducción en el entorno digital (Translation in the Digital Environment)
Universidad Pontificia Comillas	Private	Grado en Traducción e Interpretación (Four-year degree in Translation)	1. Informática Aplicada a la Traducción (IT Applied to Translation) 2. Nuevas Tecnologías Aplicadas a la Traducción (New Technologies Applied to Translation)
Universidad Antonio de Nebrija	Private	Grado en Traducción (Four-year degree in Translation)	1. Documentación y Técnicas Informáticas Aplicadas a la Traducción (Documentation and IT Techniques Applied to Translation)
Universidad de Las Palmas de Gran Canaria	Public	Grado en Traducción e Interpretación (Four-year degree in Translation and Interpreting)	1. Informática (IT)
Universidad de Málaga	Public	Grado en Traducción e Interpretación (Four-year degree in Translation and Interpreting)	1. Recursos Informáticos Aplicados a la Traducción e Interpretación (IT Resources Applied to Translation and Interpreting)
Universidad de Murcia	Public	Grado en Traducción e Interpretación (Four-year degree in Translation and Interpreting)	1. Informática Aplicada a la Traducción I (IT Applied to Translation I) 2. Informática Aplicada a la Traducción II (IT Applied to Translation II)
Universidade de Vigo	Public	Grado en Traducción e Interpretación (Four-year degree in Translation and Interpreting)	1. Herramientas para la Traducción y la Interpretación I: Informática (Tools for Translation and Interpreting I: IT) 2. Traducción en los Medios Audiovisuales (Translation in the AV Media) 3. Herramientas para la Traducción e Interpretación: Informática Avanzada (Tools for Translation and Interpreting: Advanced IT)
Universidad de Salamanca	Public	Grado en Traducción e Interpretación (Four-year degree in Translation and Interpreting)	1. Informática Básica (Basic IT) 2. Recursos Tecnológicos para la Traducción (Technological Resources for Translation) 3. Localización (Localization)

Universidad de Valladolid	Public	Grado en Traducción e Interpretación (Four-year degree in Translation and Interpreting)	1. Informática Aplicada a la Traducción (IT Applied to Translation) 2. Localización (Localization) 3. TIC para la Traducción (IT Applied to Translation)
Universidad Pablo de Olavide	Public	Grado en Traducción e Interpretación (Four-year degree in Translation and Interpreting)	1. Informática Aplicada a la Traducción I (IT Applied to Translation I) 2. Informática Aplicada a la Traducción II (IT Applied to Translation II)
Universitat de València	Public	Grado en Traducción y Mediación Interlingüística (Four-year degree in Translation and Linguistic Mediation)	1. TIC Aplicades a la Traducció (IT Applied to Translation)
Universidad San Jorge	Private	Grado en Traducción y Comunicación Intercultural (Four-year degree in Translation and Intercultural Communication)	1. Tecnologías Aplicadas a la Traducción (Technologies Applied to Translation)

Table 2: Training offered related to video game localization at Spanish universities.

It seems obvious that in most generic courses, teachers do not have the time to offer a theoretical and/or practical review with regards to video game localization content, so we looked for different data according to the course category:

- Regarding the courses which are not directly linked and/or specifically devoted to localization, we first studied the course curriculum and then contacted all the faculty members involved by e-mail to inquire about the potential educational proposals related to video game localization which could be being included as a supplementary activity within the course but not be reflected in the course curriculum.
- For those courses directly linked and/or specifically devoted to localization (comprising five courses out the total number of *Grados* related to the translation discipline offered by Spanish Universities), we observed whether or not the course curriculum included any training in video game localization and also asked the faculty members if they were implementing any activities or new initiatives which could be not updated for the aforementioned curriculum.

After studying the course curricula and checking them against the answers received from the course coordinators and faculty members teaching the subjects, we can now give a broad outline of the educational activities related to video game localization which have been implemented in the curricula for the new *Grados* in Translation. This is detailed under the following two subheadings:

1. Courses directly linked and/or specifically devoted to localization
   - There are five courses whose main topic is localization: four of these are totally devoted to this subject area and one course includes training in localization alongside scientific and technical translation (in Spanish universities, the latter two subjects usually form a combined discipline within translation training).
   - Within these courses, students receive a software localization training module as a part of the course (some only include software localization training, others offer only website localization and one combines both subject areas).
   - A number of the modules within the curricula of a few of these localization courses refer to video game localization: format, genres, language, components (software/help/documentation/other materials). Although these modules do not include a comprehensive video game localization course, they help to raise the interest of students and allow them to become acquainted with video game tools and content.
2. Courses not directly linked nor specifically devoted to localization, but which include educational content related to video game localization

Regarding the courses which include some video game localization activities but which are not directly related to this subject area, it should be emphasized that the content which does relate to video games localization is mainly sporadic; it is not set as part of the syllabus. Content emerges as short innovative educational activities (seminars, one-off classes) organized by the faculty members specializing in video game localization.

As a result of our research, we consider that overall, the video game localization training on offer is relatively limited compared to the relevance of the field of localization to the global translation business. It has been observed that partial or introductory training is preferable if the student is to be able to specialize later (Vela, 2011). Some researchers do not agree with waiting until postgraduate courses to incorporate content related to video game localization (Pérez, 2010), asserting that the industry is employing more and more translation graduates and that it is an expanding business. In our study, we therefore take this curricula overview as a starting point in order to suggest some alternative ways of introducing video game localization training which adapt to each particular situation and could meet the preferences of each faculty teacher of every course included in the new *Grados*.

6. A Survey of Professionals Working in the Video Game Localization Sector

In order to gain a clear view about the kind of training and education that today's professionals working in the video game localization sector have received, we constructed a survey with the following questions, divided into three sections:

- First Section: Studies.
  1. From which university did you graduate?
  2. What is your degree in?
- Second Section: Training.
  3. Have you attended courses in video game localization?
  4. If the answer to the previous question was 'no', has the content of any other course helped you in your professional life?
  5. Do you have additional training in video game localization?
- Third Section: Professional Profile.
  6. What is your professional profile?
  7. Do you work in-house or as a freelance?

The Survey Monkey service was used to survey the professionals[5]. It also was promoted using Twitter, Facebook and via mailing lists of translation and localization professionals. A total of 30 answers were collected. The results obtained are detailed below.

5 <http://www.surveymonkey.com/s/K62QY3T>.

6.1 *Question One: From which university did you graduate?*

The objectives of this question were to establish whether the professionals working in the video game localization sector are graduating from certain specific universities and whether they come from public or private institutions. The results are shown in Figure 3. They show an equilibrated distribution in the professionals' university of origin. Furthermore, while some universities have a greater number of people working in the sector (Universidad de Granada and Universidad Autónoma de Barcelona), the sample is too small for us to be able to claim that the programs they offer are a key element for consideration. With regards to the type of institution (either public or private), the answers reflect the natural distribution of universities teaching Translation and Interpreting Studies. Most are public, and it is therefore logical that most of the professionals working in the video game localization sector come from public institutions.

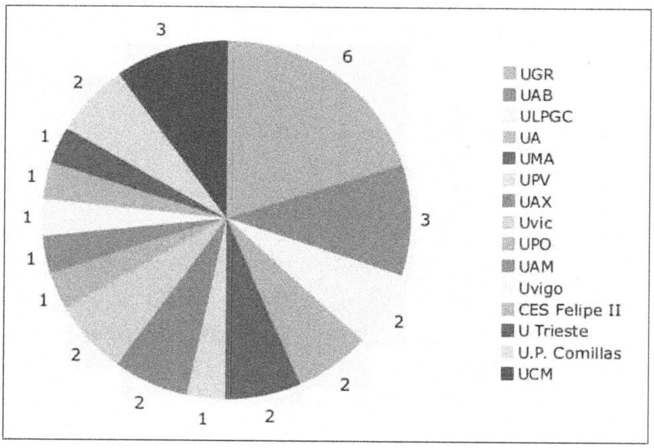

Figure 3: Chart showing survey respondents' university of origin.

6.2 *Question Two: What is your degree in?*

The objective of this question was to establish what those professionals currently working in the video game localization market had chosen to study. According to the results displayed in Figure 4, it seems that the logical way of accessing the market is through studies in Translation and Interpreting.

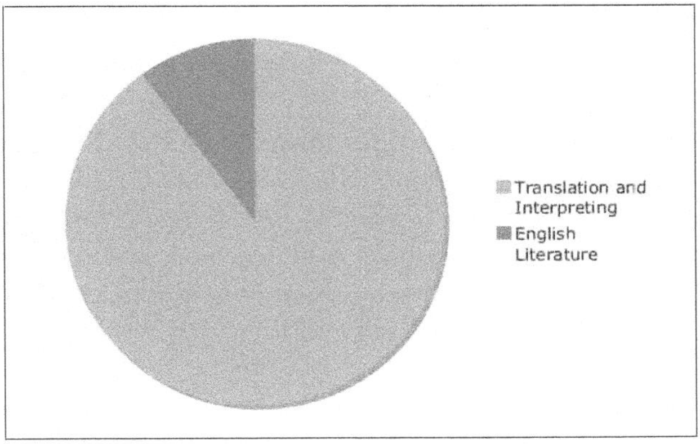

Figure 4: Chart showing the degree course attended by survey respondents.

6.3 Question Three: Have you attended courses in video game localization?

This question aimed to confirm our hypothesis that universities are not offering specialized courses in video game localization training at an undergraduate level. We therefore highlight the fact that most of the professionals who responded have received on-the-job training. According to the results shown in Figure 5, we can confirm this assumption, as most people polled answered 'No'.

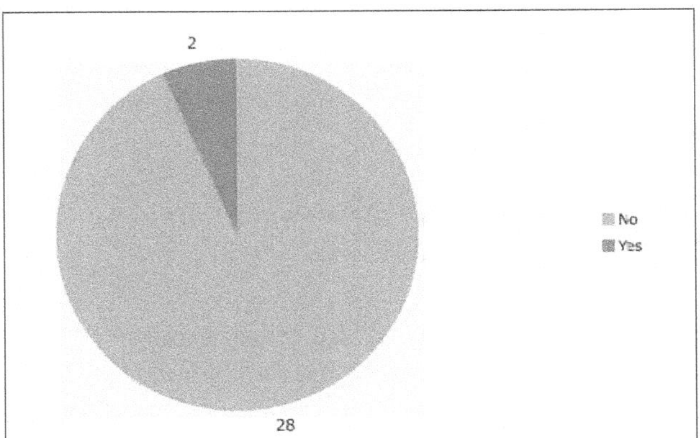

Figure 5: Chart showing the number of respondents who have attended courses in video game localization.

6.4 *Question Four: If the answer to the previous question was 'no', has the content of any other course helped you in your professional life?*

While in most of the cases specific training is not offered, we believed that other courses may provide some of the knowledge required for a professional working in the field. According to the results in Figure 6, 9 professionals have not found the training they received useful, while 21 have found several types of courses useful, mainly in the fields of software localization and technical, audiovisual and literary translation.

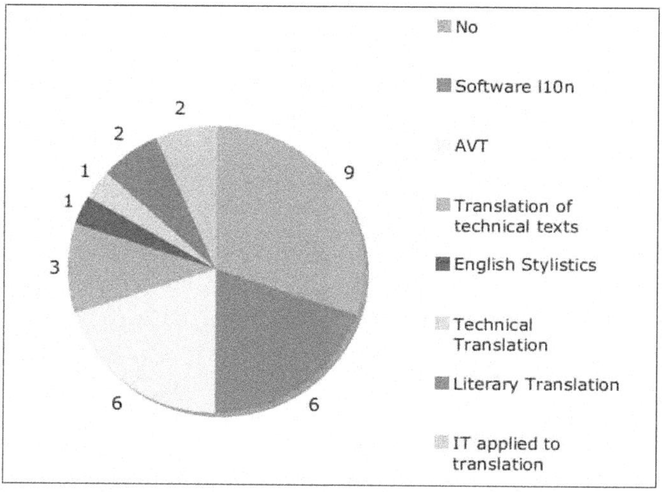

Figure 6: Chart showing the number of respondents who have found courses they attended useful in their professional practice.

6.5 *Question Five: Do you have additional training in video game localization?*

Our original hypothesis was that if the training provided by universities was insufficient, professionals would have to obtain this kind of knowledge through other means. According to the results shown in Figure 7, a significant number of professionals (15) have not attended any training program. The rest (15), have followed specialized courses, completed masters programs or received internal training within their companies.

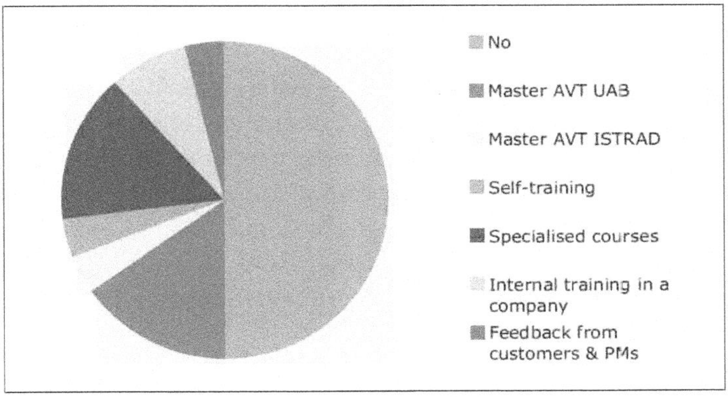

Figure 7: Additional training undertaken by respondents.

6.6 *Question Six: What is your professional profile?*

This question was oriented towards obtaining a snapshot of the roles carried out by the professionals. Most work as translators (22) or localizers (21), while only a few work as testers (6), proofreaders (1) and teachers (1). Please note that figure 8 does not reflect the total number of professionals, but rather the number of answers for each profile.

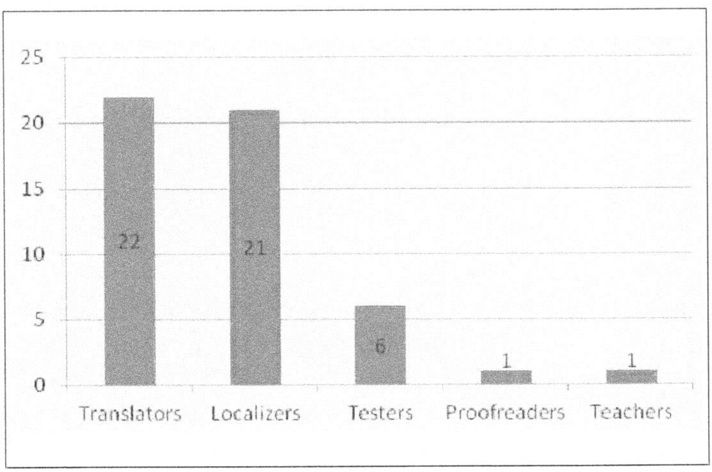

Figure 8: Professional profiles of respondents.

6.7 Question Seven: Do you work in-house or as a freelance?

This question is a continuation of the previous one. We wanted to know this information in order to develop possible didactic programs in one direction or another. For example, if more people work in-house, we can assume that some content will be learnt as part of their professional practice. However, if they are freelance professionals, the amount of content should be increased during the training period. As is shown in Figure 9, the majority of the professionals who responded work as freelancers, with only a few holding in-house positions.

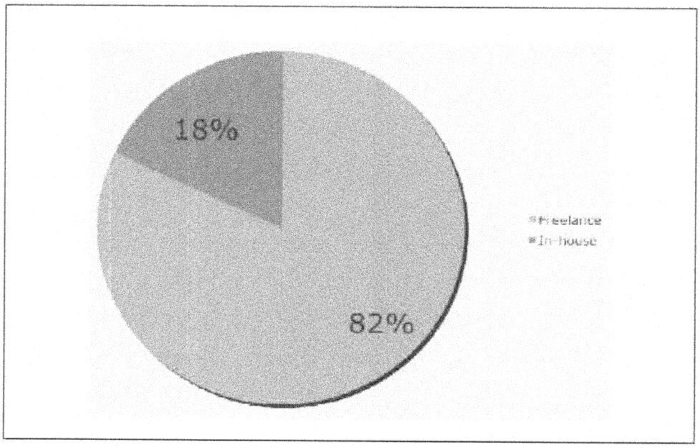

Figure 9: Chart showing the proportion of respondents working freelance and in in-house positions.

7. Conclusions

After analysing the above results, we can conclude that the importance to the work of professionals in the video games industry of the educational content in Spanish universities offered at a undergraduate level and related to video games is fairly low, since we cannot find a single undergraduate course specifically oriented to video game localization (either directly or indirectly). Considering the importance of the video games market, this asymmetry between the academic and professional worlds surprises us.

We believe it would be useful to develop a strategy for an immediate or progressive implementation of these kinds of specialized modules, according to the characteristics and limitations of each educational institution. We are aware that making changes to curricula and educational programs is always a complex processes and for this reason we propose the following levels of implementation, which may be applied either individually or in parallel:

– *An individual teaching level*: those faculty members in charge of courses not directly related to video game localization may develop what we call *integrated didactic proposals* (IDPs). These proposals are a flexible approach for integrating video game localization training at different levels. With IDPs, teachers incorporate video game localization content into other courses. For example, it would be possible to teach Literary Translation by translating a video game about Shakespeare and the Elizabethan Era, or Audiovisual Translation by teaching students how to subtitle a video game (instead of a film or series). We consider that there is a reasonable number of courses as part of which IDPs could be used.
– *Faculty staff group level*: if IDPs are successful, the teacher or teachers in charge of a General Localization course may coordinate in order to cover the different specializations inside the field. These would be Software, Web and Video Game Localization.
– *School level*: if IDPs and specific courses in Localization are well received by students, it would be possible to develop a specific major or specialised itinerary in video game localization. Here, we would include a number of courses with a progressive level of difficulty. For example, IT for Translators > Advanced IT/IT for Translators II > Audiovisual Translation > Localization > Video game Localization.

We would like to highlight a number of reasons for the importance of IDPs within this strategy:

– IDPs are easy to implement: these integrated proposals can be developed at the same time that the program descriptions for the courses are designed, with no future extra work expected for the current course curricula.
– IDPs are low cost: these proposals require neither a great deal of time nor many supplementary resources such as classrooms, computers or

additional materials, as in most cases the teaching institutions already have the necessary equipment for this type of training.

- IDPs are feasible: there appear to be are no major obstacles to developing these proposals. They are relatively easy to set up and impact quite positively on students' education, mainly due to their integrated nature – one of the main foundations of the European Higher Education Area and the Bologna Process.
- IDPs make a broader training possible: if there is no specific course in video game localization at a certain institution, these proposals at least give students some basic knowledge of the field.

Having carried out this research, our objective is now to invite teachers working with the new Translation and Interpreting curricula to consider and develop these IDPs (or others developed by themselves) in their classes, as we strongly believe that they are an optimal way of offering a form of training that is not currently present in the Spanish university system and which may become essential in order to fulfill the requirements of the market. Furthermore, while we are aware that some Spanish universities offer specific masters programs oriented to video game localization, it would, in our opinion, be wise to include at least some basic content about this field at an undergraduate level, in order to increase students' exposure to a greater number of professional specializations, especially considering the fact that this is an expanding field.

References

Activision Blizzard. 2009. 'Annual Report'. Online at: <http://files.shareholder.com/downloads/ACTI/1191355066x0x378110/4212eb1e-fd74-47f7-8abf-2de6ad88b9d3/ATVI_2009_Annual_Report_-_FINAL2.pdf> (consulted 9.03.2011).

Activision Blizzard. 2008. 'Annual Report'. Online at: <http://files.shareholder.com/downloads/ACTI/1191355066x0x219409/95c866fa-e1c7-46e9-841d-aa4ece0c4794/Activision_08_AR_full.pdf> (consulted 9.03.2011).

Activision Blizzard. 2007. 'Annual Report'. Online at: <http://files.shareholder.com/downloads/ACTI/1191355066x0x126721/44311261-a39e-43e6-b511-4cdda5308295/ATVI07AR.pdf> (consulted 9.03.2011).

Activision Blizzard. 2006. 'Annual Report'. Online at: <http://files.shareholder.com/downloads/ACTI/1191355066x0x49430/7746940C-47DC-4EE8-819A-A2366B6DDB60/ATVI2006AR.pdf> (consulted 9.03.2011).

Activision Blizzard. 2005. 'Annual Report'. Online at: <http://files.shareholder.com/downloads/ACTI/1191355335x0x25717/C14CA2C0-B130-4716-A4EA-E7CD3BA7D0AF/ar_2005.pdf> (consulted 9.03.2011).

Activision Blizzard. 2004. 'Annual Report'. Online at: <http://files.shareholder.com/downloads/ACTI/1191355335x0x25716/FA04C4A7-7347-47E3-9D7A-D3DD1B54F31A/ar_2004.pdf> (consulted 9.03.2011).

Activision Blizzard 2003 Annual Report. Online at: <http://files.shareholder.com/downloads/ACTI/1191355335x0x25715/6D28644E-16FB-4C67-9A7C-16225C465E2F/ar_2003.pdf> (consulted 9.03.2011).

ANECA. 2005. *Libro blanco. Título de grado en traducción e interpretación*. Online at: <http://www.aneca.es/var/media/150288/libroblanco_traduc_def.pdf (consulted 7.02.2011)

Austermühl, Frank. 2003. 'Training translators to localize' in *Localization and Translation Training*. Online at: <http://isg.urv.es/seminars/2003_localization_online/austermuehl.html> (consulted 9.03.2011).

Electronic Arts Inc. 2010. 'Fiscal Year 2010 Proxy Statement and Annual Report'. Online at: <http://files.shareholder.com/downloads/ERTS/1191362561x0x302630/A08800FF-9CE4-4C94-B93B-E43D90B2D195/ElectronicArts061209.pdf> (consulted 9.03.2011).

Electronic Arts Inc. 2009. 'Fiscal Year 2009 Proxy Statement and Annual Report'. Online at: <http://files.shareholder.com/downloads/ERTS/1191362561x0x302630/A08800FF-9CE4-4C94-B93B-E43D90B2D195/ElectronicArts061209.pdf> (consulted 9.03.2011).

Electronic Arts Inc. 2008. 'Notice of 2008 Annual Meeting of Stockholders Proxy Statement Annual Report on Form 10-K for the Fiscal Year Ended March 31, 2008'. Online at: <http://files.shareholder.com/downloads/ERTS/1191362561x0x302374/2EFBA688-1E2D-4857-A41D-FD4B6BA9FCA1/2008_Annual_Report_and_Proxy_Statement.pdf> (consulted 9.03.2011).

Electronic Arts Inc. 2007. '2007 Proxy Statement and Annual Report'. Online at: >http://files.shareholder.com/downloads/ERTS/1191362561x0x203377/3AD2F24E-6883-4F06-A5C2-7A5FED0E1AC9/2007%20Annual%20Report%20and%20Proxy%20Statement.pdf> (consulted 9.03.2011).

Electronic Arts Inc. 2006. '2006 Proxy Statement and Annual Report'. Online at: <http://files.shareholder.com/downloads/ERTS/1191362561x0x203378/25B063C8-94BC-4857-B821-40E2403C9829/2006%20Annual%20Report%20and%20Proxy%20Statement.pdf> (consulted 9.03.2011).

Electronic Arts Inc. 2005. 'Notice of 2005 Annual Meeting Proxy Statement and 2005 Annual Report'. Online at: <http://files.shareholder.com/downloads/ERTS/1191362561x0x203379/2D7DAE1E-C6E6-46B3-82F3-96EB084A3523/2005%20Annual%20Report%20and%20Proxy%20Statement.pdf> (consulted 9.03.2013).

Electronic Arts Inc. 2004. '2004 Notice of Annual Stockholders Meeting and Proxy Statement'. Online at: <http://files.shareholder.com/downloads/ERTS/1191362561x0x203459/404BF32B-86D6-4568-9EB7-6355E8AC4030/2004%20Proxy%20Statement.pdf> (consulted 9.03.2013).

Electronic Arts Inc. 2003. '2003 Notice of Annual Stockholders Meeting and Proxy Statement'. Online at: <http://files.shareholder.com/downloads/ERTS/1191362561x0x203461/6E0DF6E0-9A52-4C9B-9B97-09741D445DBD/2003%20Proxy%20Statement.pdf> (consulted 9.03.2011).

Kettunen, Juha and Kantola, Mauri 2006. The Implementation of the Bologna Process' in Tertiary. Education and Management, 12(3): 174–195.

Konami. 2010. 'Consolidated Financial Results for the Year Ended March 31, 2010'. Online at: <http://www.konami.co.jp/zaimu/1003/english.pdf> (consulted 9.03.2011).

Konami. 2009. 'Consolidated Financial Results for the Year Ended March 31, 2009'. Online at <http://www.konami.co.jp/zaimu/0903/english.pdf> (consulted 9.03.2011).

Konami. 2008. 'Consolidated Financial Results for the Year Ended March 31, 2008'. Online at: <http://www.konami.co.jp/zaimu/0803/english.pdf> (consulted 9.03.2011).

Konami. 2007. 'Consolidated Financial Results for the Year Ended March 31, 2007'. Online at: <http://www.konami.co.jp/zaimu/0703/english.pdf> (consulted 9.03.2011).

Konami. 2005. 'Annual Report'. Online at: <http://www.konami.co.jp/en/ir/ir-data/annualreport/report2005/pdf/all.pdf> (consulted 9.03.2011).

Lommel, Arle and Ray, Rebecca. 2007. *LISA Globalization Industry Primer*. Romainmótier, Suiza: LISA.

Microsoft. 2010. 'Annual Report'. Online at: <http://cid-4910e8dd2e872bb2.office.live.com/view.aspx/FY2010/Microsoft%202010%20Annual%20Report.docx> (consulted 9.03.2011).

Microsoft. 2009. 'Annual Report'. Online at: <http://www.microsoft.com/investor/reports/ar09/downloads/MS_2009_AR.doc> (consulted 9.03.2011).

Microsoft. 2008. 'Annual Report.' Online at: <http://www.microsoft.com/investor/reports/ar08/downloads/MS_2008_AR.docx> (consulted 9.03.2011).

Microsoft. 2007. 'Annual Report'. Online at: <http://www.microsoft.com/investor/reports/ar07/downloads/MS_2007_AR.doc> (consulted 9.3.2011).

Microsoft. 2006. 'Annual Report'. Online at: <http://www.microsoft.com/investor/reports/ar06/downloads/MS_2006_AR.doc> (consulted 9.3.2011).

Microsoft. 2005. 'Annual Report'. Online at: <http://www.microsoft.com/investor/reports/ar05/downloads/MS_2005_AR.doc> (consulted 9.3.2011).

Microsoft. 2004. 'Annual Report'. Online at: <http://www.microsoft.com/investor/reports/ar04/downloads/MSAR_10K_091404.doc. (consulted 9.3.2011).

Nintendo. 2010. 'Annual Report'. Online at: <http://www.nintendo.co.jp/ir/pdf/2010/annual1003e.pdf> (consulted 9.3.2011).

Nintendo. 2009. 'Annual Report'. Online at: <http://www.nintendo.co.jp/ir/pdf/2009/annual0903e.pdf> (consulted 9.3.2011).

Nintendo. 2008. 'Annual Report'. Online at: <http://www.nintendo.co.jp/ir/pdf/2008/annual0803e.pdf> (consulted 9.3.2011).

Nintendo. 2007. 'Annual Report'. Online at: <http://www.nintendo.com/corp/report/FY07FinancialResults.pdf> (consulted 9.3.2011).

Nintendo. 2006. 'Annual Report'. Online at: <http://www.nintendo.com/corp/report/06AnnualReport.pdf> (consulted 9.3.2011).

Nintendo. 2005. 'Annual Report'. Online at: <http://www.nintendo.com/corp/report/NintendoAnnualReport2005.pdf> (consulted 9.3.2011).

Nintendo. 2004. 'Annual Report'. Online at: <http://www.nintendo.com/corp/report/fiscal2004.pdf> (consulted 9.3.2011).

Nintendo. 2003. 'Annual Report'. Online at: <http://www.nintendo.com/corp/report/fiscal2003.pdf> (consulted 9.3.2011).

Pérez, Lucila María. 2010. *La localización de videojuegos (inglés-español): aspectos técnicos, metodológicos y profesionales.* Tesis doctoral. Universidad de Málaga.

Sega-Sammy. 2010. 'Annual Report'. Online at: <http://www.segasammy.co.jp/english/ir/library/pdf/printing_annual/2010/e_2010_annual.pdf> (consulted 9.3.2011).

Sega-Sammy. 2009. 'Annual Report'. Online at: <http://www.segasammy.co.jp/english/ir/library/pdf/printing_annual/2009/e_2009_annual.pdf> (consulted 9.3.2011).

Sega-Sammy. 2008. 'Annual Report'. Online at: <http://www.segasammy.co.jp/english/ir/library/pdf/printing_annual/2008/e_2008_annual.pdf> (consulted 9.3.2011).

Sega-Sammy. 2007. 'Annual Report'. Online at: <http://www.segasammy.co.jp/english/ir/library/pdf/printing_annual/2007/e_2007_annual.pdf> (consulted 9.3.2011).

Sega-Sammy. 2006. 'Annual Report'. Online at: <http://www.segasammy.co.jp/english/ir/library/pdf/printing_annual/2006/e_2006_annual.pdf> (consulted 9.3.2011).

Sega-Sammy. 2005. 'Annual Report'. Online at: <http://www.segasammy.co.jp/english/ir/library/pdf/printing_annual/2005/e_2005_annual.pdf> (consulted 9.3.2011).

Sony. 2010. 'Annual Report'. Online at: <http://www.sony.net/SonyInfo/IR/financial/ar/8ido18000003dkyy-att/8ido18000003dl0u.pdf>(consulted 9.3.2011).

Sony. 2009. 'Annual Report'. Online at: <http://www.sony.net/SonyInfo/IR/financial/ar/8ido180000023g2o-att/SonyAR09-E.pdf> (consulted 9.3.2011).

Sony. 2008. 'Annual Report'. Online at: <http://www.sony.net/SonyInfo/IR/financial/ar/2008/qfhh7c00000htn6x-att/SonyAR08-E.pdf> (consulted 9.3.2011).

Sony 2007 Annual Report. Online at: <http://www.sony.net/SonyInfo/IR/financial/ar/2007/8ido180000018aei-att/SonyAR07-E.pdf> (consulted 9.3.2011).

Sony. 2006. 'Annual Report'. Online at: <http://www.sony.net/SonyInfo/IR/financial/ar/2006/qfhh7c00000akslc-att/qfhh7c00000aksmr.pdf> (consulted 9.3.2011).

Sony. 2005. 'Annual Report'. Online at: <http://www.sony.net/SonyInfo/IR/financial/ar/2005/qfhh7c000005z5so-att/SonyAR05-E.pdf> (consulted 9.3.2011).

Sony. 2004. 'Annual Report'. Online at: <http://www.sony.net/SonyInfo/IR/financial/ar/2004/qfhh7c000000g7xm-att/SonyAR04-E.pdf> (consulted 9.3.2011).

Sony. 2003. 'Annual Report'. Online at: <http://www.sony.net/SonyInfo/IR/financial/ar/qfhh7c000000g7z4-att/e_ar2003.pdf> (consulted 9.3.2011).

Square Enix. 2010. 'Annual Report'. Online at: <http://www.square-enix.com/eng/pdf/ar_20100730_01en.pdf> (consulted 9.03.2011).

Square Enix. 2009. 'Annual Report'. Online at: <http://www.square-enix.com/eng/pdf/ar_20090821_01.pdf> (consulted 9.03.2011).

Square Enix. 2008. 'Annual Report'. Online at: <http://www.square-enix.com/eng/pdf/ar_20080808_01.pdf> (consulted 9.03.2011).

Square Enix. 2007. 'Annual Report'. Online at: <http://www.square-enix.com/eng/pdf/ar_20070831_01.pdf> (consulted 9.03.2011).

Square Enix. 2006. 'Annual Report'. Online at: <http://www.square-enix.com/eng/pdf/ar_20060901_01.pdf> (consulted 9.03.2011).

Square Enix.'2005 Annual Report'. Online at: <http://www.square-enix.com/eng/pdf/ar_20050831_01.pdf> (consulted 9.03.2011).

Square Enix. 2004. 'Annual Report'. Online at: <http://www.square-enix.com/eng/pdf/ar_20040806_01.pdf> (consulted 9.03.2011).

Ubisoft. 2010. 'Annual Report'. Online at: <http://www.ubisoftgroup.com/gallery_files/site/270/1042/2306.pdf> (consulted 9.03.2011).

Ubisoft. 2009. 'Annual Report'. Online at: <http://www.ubisoftgroup.com/gallery_files/site/270/1042/2139.pdf> (consulted 9.03.2011).

Ubisoft. 2008. 'Annual Report'. Online at: <http://www.ubisoftgroup.com/gallery_files/site/270/1042/1806.pdf> (consulted 9.03.2011).

Ubisoft. 2007. 'Annual Report'. Online at: <http://www.ubisoftgroup.com/gallery_files/site/270/1042/1569.pdf> (consulted 9.03.2011).

Ubisoft. 2006. 'Annual Report'. Online at: <http://www.ubisoftgroup.com/gallery_files/site/270/358/1050.pdf> (consulted 9.03.2011).

Ubisoft. 2005. 'Annual Report'. Online at: <http://www.ubisoftgroup.com/gallery_files/site/270/358/885.pdf> (consulted 9.03.2011).

Ubisoft. 2004. 'Annual Report'. Online at: <http://www.ubisoftgroup.com/gallery_files/site/270/1042/1937.pdf> (consulted 9.03.2011).

Ubisoft. 2003. 'Annual Report'. Online at: <http://www.ubisoftgroup.com/gallery_files/site/270/1042/1694.pdf> (consulted 9.03.2011).

Víctor Alonso Lion

New Challenges in Interactive Media Localization Projects

Globalization, new technology and the rise of social networks and virtual communities have undoubtedly opened up new opportunities, but they also bring new challenges to localization professionals. There must be an awareness of international requirements at an early stage of a game's creation. Internationalization of code and content has to be considered from the very beginning of the design process. New online project management and collaboration tools, combined with easier outsourcing, make possible the creation of global teams for companies of all sizes. The amount of content requiring localization has increased significantly. Developers and distributors need to consider alternative approaches for different types of content. A direct translation is not a valid method of engaging customers. Players need to feel their needs are taken into consideration in the games they choose to play. Localizers and developers must work together in order to produce a set of products that will please the different markets. Localizers need to be proactive and propose changes, while developers need to code while keeping in mind that their game will need local adaptations. This article explores the trends mentioned and tries to provide guidance on both successful and unsuccessful methods for approaching game localization projects in a new global, community driven and locally-focused environment.

1. Introduction

Interactive media and games localization is a particular field in and of itself. New discoveries and new solutions need to be put in place each and every day. All professions working in this field (encompassing a diverse range of specialties) need to explore and find solutions to new challenges in interactive media localization projects.

Localization professionals who do not work in interactive media see localization as an obscure and mysterious process. Many unfounded accusations arise when outsiders talk about game localization. The usual resultant topic areas are: lack of information, lack of standardisation, lack of procedures, lack of planning, low quality, few academic and professional references and so on.

In fact, the reality is different and the game localization industry is at least as mature as that of standard software localization. There are naturally many differences and maturity levels depending on the particular team, company, developer or distributor. For each game, each and every party interacts in a different way. The way in which these parties are organized and interact with each other is completely different from one game to another.

The fact is that game localization faces the most complex challenge: localization of continuous innovation. This begs the question as to why the game localization industry needs to pay special attention to this continuous innovation. A long list of factors needs to be considered when managing and localizing games: genre, game type, platforms, game engine and of course the process currently in place. In addition, factors such as how teams are constructed and how the multiple parties are organized are important. In the game industry, these two particular aspects could be quite complex and diverse. Games are actually hard to define and it is difficult to consider them as a single and easy to describe product.

The purpose of this article is to attempt to explain the environment of continuous change nowadays faced by localization. Furthermore, there is a good chance that this change will continue and increase exponentially.

In fact, as the author sees it, these changes are a great opportunity for those willing to reconsider their processes, those who are ready to reconsider what is the "right way" of doing things, and those who are open to investigating new ideas and transforming these in new opportunities.

Those individuals and companies who are open to the idea of "pressing the reset button" on occasion are likely to be most adaptable to change. The field of interactive media and video games particularly offers a framework which is consistently at the cutting edge of new technology. Team members dealing with game localization are those who will have the opportunity to test new processes and tools.

So what is occurring in this environment that results in having to face so many substantial changes? What exactly is this strange ecosystem of new media localization in which we are currently working? First, this paper will attempt to describe this ecosystem and a number of the enablers/facilitators of change.

2. The Social Media Landscape

Using today's technology, we have collectively created a layer of worldwide interactions. Some call this layer "social media", but we could probably simply consider it to be the rise of new communication channels[1]. Video games, social games and virtual worlds all play an important part in this. The interaction does not need to be pre-programmed and can be triggered by the community itself. Returning specifically to games, there is an increased interest and growth in an additional "gaming layer" for all these interactions. Games therefore play an important role in the opening up of communication channels between creators and users.

This is evident in the rise of crowdsourcing initiatives and increased community participation. Other industries have a profound interest in this "gaming layer". It appears very much as if education and eLearning will be the first fields to join this "gamification" approach, as there are already multiple initiatives in educational institutions and big corporations which adopt this kind of motivational methodology to help students and employees to get engaged and involved in the learning processes.

3. The Global Increase of Technological Access

Cheaper access to the internet and the resultant increase in technological access[2] is a second enabler which will certainly facilitate this "revolution". The rise in broadband connectivity will definitively provide access and open collaboration channels to millions who are not even considered to-

1 Schell J. 2010. Ted.com presentation "When games invade real life": <http://www.ted.com/talks/jesse_schell_when_games_invade_real_life.html>.

2 <http://www.morganstanley.com/institutional/techresearch/>.

day. This will not only bring significant changes to content, but also new competition across all areas.

4. The Global Playground

It is of course no longer news that we live in a globalized world[3]. Interactions from Beijing to San Francisco, Berlin to Barcelona or Singapore to Dublin occur in seconds. This is of course a general trend in all content-related industries, and there is nothing specific to localization professionals or professionals working on interactive media and games. However, it is still a radical change that is important to note. The importance of a specific market can rapidly shift, and a target language that was essential at a certain point in time or for a particular game, may not be that important for another product on a different date. These changes occur rapidly. For example, some very popular current markets for online social games include Indonesia and Turkey, ignored by many just a couple of years ago. There are of course great expectations for the future for markets like India, where broadband penetration is increasing exponentially with the natural result of greater numbers of players of all sorts of games.

There is no doubt that the global landscape is altering. But the processes, technology and professionals working on the new projects are changing too.

Anyone interested in remaining in this industry should be very adaptable to change. The statement that one game localization project is similar to another is difficult to maintain.

All game projects seem to be different. Players want and need to be challenged by new technology, new adventures and new art. They constantly want to live new experiences and to feel increased sensations of intense emotion each and every time. We cannot forget that a big difference between games and other product types is that users and consumers of video games have explicitly chosen this particular product for their leisure time.

3 Friedman T. 2005. *The World is Flat.* Farrar, Straus and Giroux Ghemawat, P. 2011. *World 3.0: Global Prosperity And How To Achieve It.* Mcgraw-hill.

This situation involving a level of instability and uncertainty influences the way all interactions within the industry occur: both amongst professional teams working on the project, and in the feedback that comes from all the project stakeholders, including the end-users.

Teams need to adapt to the new game engines, to the new content management systems, and especially to all the new unexpected requirements. This environment of continuous change must be considered an opportunity for growth, and an opportunity to investigate and learn new approaches. Change is a small price to pay for being part of such a revolutionary industry.

5. Process, Templates and Standards

Is it at all possible to use templates for such projects? Can you really reuse plans and budget information etc.? Can you leverage the experience gleaned from a former game? These questions aside, there is no doubt that "flexibility" and "adaptation" are becoming the most important skills in game localization.

There are of course opportunities for recycling ideas and experiences. The knowledge acquired in past projects must be leveraged to increase the level of service and professionalism delivered by the localization teams involved in a project. Teams need to learn from their errors, and need to use the information to create knowledge databases concerning risks they have mitigated and surpassed. Risk management plans and "lessons learned" documentation prove to be essential.

The essence of the change that is present in game localization projects already exists in the core of the product. The engines used to build games continuously change and evolve. Any investigation of the different commercial engines publically available will uncover hundreds of possible options. In addition, we can safely say that many AAA titles (which are essentially high-quality games with large budgets) and MMO titles (massively multiplayer online games) develop their own engines from scratch. This is an excellent opportunity for providing advice on internationalization and for professionals capable of implementing the requirements needed to efficiently localize the products, thus helping to minimize the cost of localization of these games.

With regards to standardization, we have to acknowledge the fact that creating and applying standards in such an environment is very difficult. Standards in the area of the development of games are constantly changing. Applying localization standards to this kind of product is a difficult, and at the very least challenging exercise.

So what should the standards for game localization be?

At this moment in time, excel files, text files, comma separated value files, and fortunately now formats like XML, are the most popular formats used for game localization. However, these are collectively nothing other than the most common formats. This is very different to a true *standard.* Some popular localization standards such as XLIFF, TBX or TMX will definitely be used in the future, but these standards are for the localization side only. As localizers, we need to be able to process the content we receive and then use localization tools and adapt them for each project in order to finally be able to deliver what the developers are requesting in the different target languages. Trying to convince developers to learn and apply localization standards that do not in fact particularly concern them will probably in most instances be a futile exercise. So why is this happening? For development teams, the formats used in requests for localization services provide the safest option for exporting and importing content needing to be localized. Furthermore, these formats are the closest database replicators where information and metadata essential to translating the content are present for the professionals adapting the content to another language (at least for xls and csv files).

The point of view of the developers is that presenting games for localisation in these formats is the only way to ensure that translators do not mess up the code. In addition, it is a way to avoid spending a huge amount of time copying and pasting translated content across. One could ask whether there is a way to standardize engines. Is there a way to impose the use of standards which are compatible with translation tools? The reality is that the success of interactive entertainment cannot be programmed, and can hardly be predicted. The design of every game is different and code is different in almost every project. Forcing a standard would not work for such a creative and diverse field.

Some advocate evangelization. Naturally, some stable teams can learn and improve and put in place best practices. These best practices are undoubtedly useful and vital. However, localization professionals will still continue to face many of the problems that to them seem old and familiar but are completely new to the continuous flow of new people working in the game development industry. Game localizers need to acknowledge the fact that a large element of their work involves explaining and teaching their needs and requirements to new teams. Not only that, but they also need to be able to communicate the idea that following their advice will benefit the companies wanting to localize their games. In order to be heard and listened to, localizers must, in a concrete way, be able to show that good practices will benefit the team confronted with the responsibility of "mixing" content and code. A good analogy for this would be to consider good localization professionals as teachers needing to find a way of transmitting this knowledge to a new team of pupils for each and every project. The basics will be the same, but the best practices will evolve and the knowledge and communication style will need to be adapted to the different teams with whom these professional localisers are working.

Additional challenges are everywhere. It is not only the development teams, but also the other stakeholders who require education in the needs of localization: from local marketing teams wishing to express their points of view to the community that will actually play the games (the amateur scene has increasing influence and means of expressing its views).

Slightly different from the specific localization standards and tools is the challenge of and discussion surrounding content strategy. Content strategy, considered simply as the means of organizing content, the tools related to it and its architecture, is something that may well sound appealing to some of the development teams with whom localization professionals need to work. Since development teams can face interesting and even challenging issues surrounding the organization and management of content, providing help and valuable advice in this area can, in return, help to make the localization process easier to manage and the content less difficult to work with. The language of content management systems (CMS) and global content management systems (GCMS) is something that development teams can understand, whereas the specific language relating to linguistic issues is normally not.

As a last word in this section, I want to insist on the fact that whilst templates and best practices are crucial, they are nothing but a foundation on which localization professionals can build their communications and added value services. There is no doubt that in large projects, this documentation and acquired knowledge will be of use. In short, having this essential knowledge and the processes clearly set out will help localization professionals to communicate this information with increased effectiveness and less effort each time.

In a way, localization professionals need to offer this stability and added value to the many new teams and individuals facing the challenge of localization with whom they will need to cooperate.

6. Flexibility

In game localization, flexibility is more than a simple buzz word. Where there is content, there are changes. Examples of flexibility requirements are the manner in which multiple projects no longer arrive as a single large group of files needing to be dealt with and organised before planning the translation, but instead are now divided into batches of smaller translation requests.

The localization process needs to be ready to allow a continuous production flow based on iterations. Making a project management analogy, a waterfall project (essentially a step-by-step process, where a stage needs to be finalised in order to proceed with the subsequent one) will not work. Instead, an agile approach (following the idea of getting something workable as soon as possible, where features – or in the case of game localisation, content – is added as required in order to arrive at the final product) will better suit the process.

Localization processes need to be put in place with the idea of processing content and files on a per batch basis. This must naturally occur whilst keeping in mind the product that is being created. It is therefore crucial to offer a solid methodology: there must be style guides, glossary creation and maintenance, knowledge of the product and the ability to transmit the original idea. This is equally as valid and necessary for whoever is handling the localisation process, be it a freelance translator, a team of linguists or a company. The structure and way of working has to be flexible, with a confronted and proven methodology.

In addition to the fact that localization projects do not come as a single request, we have to consider the particularity that projects are completely different in size from one to the other. Crucial in order to be able to handle game localization projects is the ability to scale production up and down, to be big or small, depending on the project and game involved. This not only applies to companies, but is also important for freelancers who need to be able to collaborate with their peers and even create small teams where required by potential clients.

7. Collaboration

We can consider translators, linguists, localization engineers and project managers working in localization as technicians. These technicians can no longer work in isolation'. There is an increased need to understand the overall requirements of clients. This includes understanding the "why" of this continuously changing environment. As we have seen, game localization professionals need to be adaptable, but most especially they must be connected and master the skill of continuous communication.

Some projects with a massive amount of content need very different specialists, even considering only the translation side of the localization equation. One lone freelancer will not be able to handle the entire range of different content types. These projects often demand a surplus of process control. Not only will projects need to be shared because of tight deadlines and high content volume, but they will also require teams to provide the right set of glossaries, the ability to discuss terminology and the willingness to agree on the right choices for style guides.

Mastering collaboration tools such as wikis, discussion forums, and any other tool that enables the sharing of files and knowledge is now mandatory. Even if some clients do not use these tools and rely on e-mail for all their interactions, there is a very good chance that they will be using these technologies in the near future. Furthermore, if discussing freelance translators, it is not only peers with whom they need to cooperate, but also the different professionals involved in the other stages of the process. Even if an attempt is made to categorize the levels of services offered (Multilingual Language Vendors (MLVs), Single Language Vendors (SLVs), freelancers, etc., the reality is that all these parties are in the same boat and will, on occasions, need to collaborate. Without doubt, collaboration and a discipline to handle this ever changing production process is needed.

8. Quality

Quality in video game localization is mainly related to the following factors: emotion, content types and engagement.

Emotion

Quality in video game localization is not only a matter of linguistic quality. Dealing with the localisation of emotion is in fact the most challenging and difficult part of the job: is it really possible to trigger the same emotional reactions in different regions of the world? Will the same jokes work in Tokyo, Brasilia, Paris, Berlin, Istanbul, Jakarta and New York?

There is a very big challenge when considering how to measure the strange concept of 'quality'. A perfect replica of the source is, of course, not often the answer. In most instances, in video games the challenge relates to creating an atmosphere similar to that of the source but still appealing to the target market.

In addition, it is important to understand that different games can trigger different emotions in different countries and contexts. The way these games sell in different cultures is also an interesting aspect to study and analyse. Cultural checks performed by experts are becoming popular and prove an important step in the process to avoid potential problems when the games are launched in different regions.

Content types

Again, we cannot consider all types of content to be the same. The user interface will need a certain type of translation, the marketing material a different one and, in most games, the storyline will require yet another different approach. This is simply an example of an arbitrary split of content, but there could be many more content types that need to be considered. Within the storyline for example, there could well be instances where a local "marketing" type of speech will be crucial, surpassing the need to be aligned with the source.

Creativity is also one of the challenges that game localization often needs to address. The methodology for transmitting the idea the source aimed to transmit is crucial. It could be true that video games are basically creative products. Yet we cannot forget that many game elements are not really creative and require different localization skills: in-game messages, legal content etc. The overall product still needs to be innovative, appealing and entertaining.

Returning to measuring quality, one of the tendencies is to measure quality by comparing how similar groups of users in different countries with different languages consider the translated content. This of course, can be achieved in the case of online games where the feedback from the community is almost immediate, allowing this comparison. The idea is to measure through the success of a game whether or not it is correctly localized. This is a dangerous exercise since many other factors could come into play, but it is still very interesting approach.

Related to the issue of quality is also the question of who must decide what constitutes quality in these kinds of entertainment products. The identification of stakeholders is one of the most intense exercises faced by those working in the industry. On the one hand we have the target market and on the other the need to comply with the requirements of the platform creators. Finally, in-country marketing teams also have their own input to the discussion.

Engagement

If we want to engage our players, we need to understand them. We need to deeply get to know them and uncover the tricks we can use that will make them love our work.

Translators need to be fantastic writers and copywriters. They need to know when to use an expression which is closer to marketing and when to employ a style which is more creative and literature-based. Direct translation would work on occasion, but in most instances it is not valid for engaging players. These players need to feel that they are taken into consideration in the games they choose to play. Translators need to adapt to the culture and locale.

Sometimes this local adaptation is even imposed by legislation. However, most of the time it is the body of game players who require quality localization. These players scrutinize every single detail of the localization output. The more popular the game, the more critics there will be. Today, this feedback has a big impact. Game fans can use social media as "loud speakers" to spread their message (positive or negative). It is crucial not to forget that users are fans and are emotionally bound to the product. They could really feel disappointed and let-down if a game translation, voice-over or functionality in a specific language was not functioning properly.

9. Transparency, Confidentiality and Security

Naturally, localization approaches vary depending on the specific needs and the global ambition of the developer and distributor. New approaches such as online project management and collaboration tools, combined with easier outsourcing, make the creation of global teams possible for companies of all sizes. An interesting and challenging consequence of this is that it is almost impossible to hide the approach of the localization process, or the identities of one's collaborators or outsourced partners. Scenarios in which the local localization team needs to work together with the local PR team of the final client, or the marketing team on a regional level, are becoming increasingly popular. The same can happen for a team of translators: they may need to work together at the client premises to test the game, or clients may require each contributor to sign a personal NDA (Non-Disclosure Agreement). Companies especially need to adapt to this new situation and to not be afraid of this new kind of environment. Control and obscurantism are not as successful in our connected world.

The security requirements of the games industry are high. Clients need to secure their assets, and localization teams and resources need to guarantee the security and confidentiality of the content assets with which they work. Some still even opt for using locked rooms in testing departments and high security standards such as the NATO Security labelling have been popular and a saleable asset for some companies. Trust plays an important role in this, both for external and internal teams. It is also common for developers to bring a large part of their production in-house, with the inevitable associated cost generated by having a complete internal international localization team.

Today, technology has provided options that comply with the requirements for security and confidentiality and still allow for the unavoidable fact that local localization teams are by their nature spread all over the world. Cloud computing and virtualization, when correctly setup, comply with these required guarantees. Clients and providers can ensure files, proprietary tools and content never leave their network or servers and still permit access to an external collaborator by setting the necessary rights. Being trained in and able to work with this type of technology will increase the chances of clients trusting localization professionals with their valuable content assets.

10. Technology

Localizers are not developers and nor do they need to be. However, it is very important to master the technology required to perform the job of meeting client requirements. Mastering some tools is simply essential in order to just be considered as a possible resource option. It is a reasonable request for a localizer to be up to date with technology. Localization teams not only need to be able to speak a language relatively close to that of the developers, but they need to master productivity tools and many of the other tools used by clients.

Needless to say, localizers need to master lingua and translation-related technology (from Computer Assisted Translation Tools to Content Management Systems). Ideally, each specialist would focus on his or her own field, working towards delivering the perfect output. Content would be organized within simple CMSs and translation tools would comprehensively provide all the necessary context information. The question is whether this ideal tool or process is still to come, or whether it is simply not possible to anticipate the needs of developers and have a tool that would work for every type of project. There is a good chance that this ideal tool would not fit with new technology employed in the next generation of games.

Localization professionals also need to master the use of communication and sharing tools. It is essential not to forget that the videogame industry is an industry which aims to be at the cutting edge of new developments. Being part of this industry as game localizers means being up-to-date on new tools, and being open to testing, trying and investing time in researching new approaches. It is challenging, but definitely worth the effort.

Furthermore, localizers need to understand how the content that needs to be translated is constructed. They need to understand the content management systems with which games are developed, and also need to understand automated language strings, on-the-fly messages etc. In short, they must be at least familiar with a few of the basics of the content engineering that is being performed by their clients.

A big advantage this brings is being able to explain to client engineers what they can and cannot do with the content if it is supposed to be localized. It is a very common occurrence that content is not correctly internationalized and unfortunately remains that way, simply because it was not possible for the localization teams to communicate their requirements and the necessities of their own process. On many occasions the changes needed in the code are simple and easy to implement. In addition, the perception of added value to the client when advice is provided is in this sense very positive and can result in the formation of long-term relationships.

Localization cannot be just an afterthought. Creating a product without anticipating its possible global reach is not really an option for the developers of an ambitious interactive media application. The first reason for this is because without preparation, the possible success in a given market will not be replicated on time in other target markets. Customers do not really wait, and the most common situation is that a new popular product is released while the unprepared team is working on making their own product international, losing crucial time in which to make their project an international success.

Localizers and developers must work together in order to produce a set of products that will please the different markets. Localizers need to be proactive and propose changes, while developers need to code keeping in mind that their game will need local adaptations. As mentioned previously, code and content internationalization has to be considered from the very beginning. An awareness of international requirements needs to be present at an early stage in game creation in order to facilitate a smooth internationalization process.

11. Content

In the early days of interactive media, there was little content to localize. Today, some game designs are based on the idea that a game needs to be simple and global and as such could have a more basic design. Developers need to please everybody with a few basic concepts. However, in the past most games were built this way (see Pong (1972) or Pacman (1980) as examples); they were games with no text content (and therefore no localization needs), except perhaps for tiny instructions and a "Game Over" statement.

The amount of content in games is impossible to estimate, as it ranges from no words at all to several million depending on the game. However, to simplify this, we could estimate that a basic social game on a social network starts from around one thousand words needing to be translated; an average AAA could have around half a million words; and large RPGs (Role Playing Games) or MMOs (Massive Multiplayer Online Games) can easily reach over 1 million words to be translated, with continuous expansion if the game is successful.

Crowdsourcing

As previously discussed, the amount of content needing to be localized has, on average, significantly increased. For some games, developers and distributors consider alternative approaches for different types of content. Wikis, forums and online guides are even sometimes managed by players themselves. This approach, if correctly used, can be a good marketing strategy and enable the creation of alternative content related to the product. However, those who consider crowdsourcing to be a cost saving approach are failing; the process of splitting the content between a large number of volunteers whilst still maintaining quality and consistency can easily extend beyond the cost of a traditional localization approach. That said, this could of course work for some products where fan participation and the use of this approach would create a sense of community and allow the creation of solid networks of contributors and buyers.

No static content

Content to localize rarely comes as a single big volume request. The need for a simultaneous shipment in multiple languages forces development teams to begin integrating source and localized content on a step-by-step basis. This is also a reflection of the development process itself: a part of the game completed equals a part of the game that needs to be ready in multiple languages. This is one of the most important characteristics of the game localization industry: the need to deal with continuous change. Not even AAAs with years of previous development are static anymore. NextGen consoles allow for DLC (downloadable content). Here, we are not only talking about MMOs which are really virtual worlds in themselves, some having new content created and localized on a daily basis. In fact, console games are nowadays no different from MMOs and permit the integration of extra content and even corrections on an ad hoc basis.

Working with this fragmented yet still massive quantity of content is a real challenge for localization teams and professionals. Project management approaches and processes have to be ready for working in iterations. The professionals responsible for organizing the work need to have a methodology on which to base their decisions, have the necessary knowledge and must prove to be proactive in order to succeed with such a challenge. The localization process needs to be prepared for receiving content in all possible ways.

12. Influencers/Community

The community has increasing influence on game content. In some games, it really appears that the community owns the content and influences both source text and translation. Members of the community also create their own content related to the games of which they are fans. An example of this is the second biggest wiki in the world, completely dedicated to the virtual world of a popular MMO (Blizzard's World of Warcraft[4]). This really is a

4 McGonigal J. 2010. Ted.com presentation "Gaming can make a better world": <http://www.ted.com/talks/jane_mcgonigal_gaming_can_make_a_better_world.html>.

new type of content that did not previously exist. In the case of this specific example, it is of course not really localized content, but content that is directly created in the target languages. It is a case of people creating more and more content providing information about a world that is not 'real'.

The community even influences the way in which localization decisions are taken. The input of the final user easily reaches the clients, the developers and the press who are discussing current and upcoming games. The community can to some extent influence what the next expansion of a game will look like, what features developers will create or what kind of characters are popular and unpopular. This input comes from the social network layer to which users have given power, and in fact users themselves have created, in the sense that social networks are shaped by participation.

The intellectual property (IP) on which the games are based is still present and of course is the basic element which triggers the community to interact. However, it is not really clear who owns the stories, ideas and end result of an interactive media product.

13. Conclusions

Game localization professionals attempt to find a "stable process" within an unstable environment. They search for a framework of continuous change, which pushes them to the limit and at all times makes them challenge a status quo which is difficult to define. The instability created by continuous change will certainly push the boundaries of what can and cannot be accomplished on different levels, and will undoubtedly lead to a real revolution in localization.

What are the strategies that enable survival in such an environment of continuous change? I personally believe that project management as a strategic competency is essential, not only for companies working on localization but also even for individuals. If we consider project management to be the discipline for addressing and navigating change, risk and opportunity, it must be suitable for the evolving environment of game localization.

Localization of interactive media, games and new media is an area of the localization industry that enables research and process improvements, and could lead to significant innovation.

Kate Edwards

Beyond Localization: An Overview of Game Culturalization

In the practice of video game design, development and distribution, the act of localizing games for international markets is well documented. A less considered aspect is that of cultural adaptation, i.e., the process of culturalization in which the game's assumptions and choices are assessed against the complex cultural landscape of local markets, including primary factors such as history, religion, ethnicity and geopolitics. By considering these deeper level aspects, game content can be better adapted to avoid the potential for negative backlash and to augment games with more locally-relevant content.

1. Introduction: Culture is Critical to Game Design

In October 2010, a controversy arose when Electronic Arts' new version of *Medal of Honor* allowed game players to assume the role of the Taliban insurgents in Afghanistan when playing in multiplayer mode. The decision quickly yielded backlash from U.S. military groups and their public supporters, citing that it was insensitive to include such an option, particularly when the U.S. was still in conflict with the Taliban. EA opted not to change the game, and others questioned why it was any different from games where the player could assume the role of a Nazi German soldier during World War II. But the public anger was coming from the point of historical proximity; World War II was well over 60 years prior while the conflict against the Taliban was contemporary and still very tangible to many people. The contextual proximity in both time (the present) and situation (U.S. soldiers fighting and dying) made the potential response much more volatile. Also, the general public perception that a game is supposed to be "fun" strongly influences their negative response; in other words, they perceived that EA was turning the present day fight against the Taliban into something fun and enjoyable.

We hear about incidents like this from time to time in the game industry as well as in other business sectors. For instance, marketing mistakes such as product names that don't translate well into other languages or marketing campaigns that are misunderstood are very common and they usually provide some amusement. However, such mistakes can prove to be costly for companies – not just the quantitative loss of potential revenue from a specific locale but the greater qualitative effects of negative public relations, damage to the company's image, and a strained relationship with the local government. In the worst scenario, a government may not only ban the product but also take direct action against the company's personnel, such as in a local subsidiary. Of course this isn't the desired result for any game publisher, regardless of how edgy they intend their content to be. Thus we must ask: how can game developers prevent problems like this from occurring?

Culturalization takes a step beyond localization, making a more fundamental examination of a game's assumptions and choices, and then assesses the viability of those creative choices in both the global, intercultural marketplace as well as in specific locales. While localization assists gamers with simply comprehending the game's content through translation, culturalization allows gamers to engage with the game's content at a potentially more meaningful level. Or conversely, culturalization ensures that gamers will not be disengaged by a piece of content that is considered incongruent or even offensive in the game's environment.

Localization is an assumed to be a regular aspect of game distribution by most game developers and publishers. North American game publishers regularly localize their games into FIGS (French, Italian, German & Spanish) and Japanese, as well as others like Chinese, Korean, Russian, and Scandinavian languages on a more regular basis. While a significant portion of the global game industry's revenue is being generated through localization, and there exists a growing international focus within the industry, localization alone will not be enough to sustain revenues and interest. This is particularly true as game development starts to increase on the local level in emerging markets, where local content will compete directly with global franchises.

Most localization efforts on a game currently occur late in the development cycle which unfortunately leaves little room for ensuring a thorough and effective localization. The better examples of game localization are usually ones where all the relevant aspects of localization were addressed throughout the project, from start to finish. When it comes to content culturalization, such considerations must be made from the start of the game's design. In order to account for locale-specific content sensitivities and look for potential opportunities, culturalization requires a *proactive* approach, wherein the global viability of the content is considered throughout development. Stated more bluntly, culturalization isn't only a required step for selling overseas; it becomes a *modus operandi* for game design, development and distribution which acknowledges the reality of the multicultural gamers who will be exposed to the content on the first day of release.

Therefore, this chapter will briefly discuss some key aspects of game content culturalization while leveraging numerous examples. It will conclude with a basic culturalization methodology that may be employed in a typical game development cycle.

2. Levels of Game Culturalization

The need for game localization is a well-known necessity within the game industry, particularly with the reality that roughly 50% of the industry's global revenue is generated from localized versions. For individual companies, the percentage of annual revenue derived from localized versions of their games can range widely, from 25% to as much as a very significant 70% (as told to this author by representatives of various game developers). Also, some companies are beginning to discover the value of retroactive localization, where they may revive an older game title that has exhausted its run in typical gaming markets and spend a nominal amount on localizing for newly emerging markets. In some cases, the return on such an investment has been staggering, as much as 400% (as related by Jaime Giné, VP of Globalization, in a keynote speech at the 2009 Localization Summit at the Game Developers Conference). As game players become somewhat saturated in typical target locales, localization has become an even more appealing avenue to maintaining and/or expanding a company's revenue stream.

So there's no question that localization is critical and necessary. However, the need for *culturalization* and taking a step beyond language remains an often overlooked yet much-needed reality. In fact, from this author's perspective, what we understand currently as "localization" is actually just one element in a broader scale of content culturalization. In its most rudimentary form, content culturalization can be viewed as the following three phases:

a. Avoid disruptive issues to allow a game to remain in the target market.
b. Perform "typical" localization to allow the game to be understood.
c. Adapt and provide locale-specific options to allow the game to be locally relevant.

Stated simply, a more straightforward way to view these three phases is as follows:

a. Reactive culturalization: Make the content viable.
b. Localization & Internationalization: Make the content legible.
c. Proactive culturalization: Make the content meaningful.

Let's examine these three phases and elaborate further. Making content "viable" essentially means allowing the content (and by implication, the product/service that contains it) to remain in the target locale. This means that there exists no issue in the content that could cause a local government and/or consumers to react negatively and thus initiate a potentially costly and embarrassing backlash episode. This step, which could be alternatively labeled as "reactive culturalization," is culturalization at its most fundamental level: simply flagging and removing potentially problematic issues from the risk equation.

A good example of the viability issue would be *Fallout 3*, released in 2009. In its post-apocalyptic landscape around Washington D.C., the game included a creature that was called a "brahmin", which was a mutated, 2-headed Brahman bull. The animal could be used for carrying loads, or it could be killed and consumed (even though it was radioactive). The presence of this single issue made the title inaccessible to the consumer market in India, because Brahman cattle are considered sacred to the Hindu religion and the government actually maintains laws that protect the animals from harm (the real ones anyway; it's unclear if such laws extend to virtual representations). Revising this one aspect would have made the game viable for the Indian market; it could have been almost any other animal and it would have been acceptable to use.

Enabling legibility is a more straightforward phase, as this is what is perceived to be common localization practices. Legibility is meant here to simply describe the player's ability to understand the text contained within the game, most usually in their preferred language. Of course, this entails the process of text and audio translation and providing proper tools for translators so that the content is better understood by local players. This also includes all the other internationalization aspects that must be implemented for local compatibility, things like field size considerations in the user interface, ensuring the use of Unicode to accommodate various scripts, and so forth.

Perhaps one of the better negative examples of the need for legibility comes from the 1991 game *Zero Wing*, in which an opening cut scene declares "All your base are belong to us." The poor translation has taken a life of its own as a meme in popular culture and has been replicated in many contexts as a classic translation error. In a more positive light, some localization efforts are monumental and very successful. In 2008, CD Projekt's Polish edition of *Mass Effect* proved to be the largest localization project in the history of Polish media (of any type of media). It received wide acclaim for its high quality, aided by famous actors from Polish cinema who contributed to the effort.

The final phase of making content more meaningful for local gamers is a step that many developers are starting to grasp and implement. It's arguably a more time-intensive and research-dependent aspect of content development yet the return on the investment can potentially be very significant. Obviously we're not talking about language here but rather about tailoring at a more fundamental level, so that gamers perceive the game as something as "local" in nature or at least very locally relevant. From a game community aspect, adding more local meaning to a game enhances the reputation of the game developer as they become known as a studio that truly cares about locally-specific expectations.

There have been several good examples of adding local meaning, in a step that could also be labeled as "positive culturalization." In some car racing games like *Forza Motorsport* (2005), care was taken to add locally-relevant automobiles for localized versions to that the gamers are enjoying the types of cars with which they're more familiar. Another example entails the card game basra, which is very popular throughout the Middle East. The game has been turned into a mobile version and then the artwork was culturalized for individual Middle Eastern locales, so the appearance of the Saudi Arabian version was very accurate for that country, as was the version for UAE and so on. This simple investment in additional artwork paid off as the company saw a very positive response and thus increased revenues.

In regards to these phases of culturalization, it's critical to elaborate on three key aspects. First, under this framework it should be clear that the common practice of localization is viewed as a component part of a broader practice of culturalization. As important as we recognize localization to be, the process of achieving legibility in other locales through translation is not the only step required in preparing content for consumption in other cultures. This is true for video games as much as it's true for every other type of content. Granted, there is a divergent perception of what "localization" entails, but the focus here is primarily on the translation aspect.

Secondly, it can be argued that a game title should be "legible" before it is "viable," because the product needs to be able to be understood by local players, correct? However, from this author's experience is has been evident that the issues which prevent a game from remaining viable almost always supersede the need for linguistic understanding. The basic logic is that a government will ban or restrict a game based on sensitive content regardless if it's localized or not. Thus before providing a game in the local language, it's critical that the game can even remain within the borders of the target market. The reality is that many games can in fact be played perfectly fine with little or no localization effort; this doesn't downplay the importance of localization but rather it emphasizes one of the unique aspects of game content: there is a logic to most games that transcends language and appeals to a broad toolkit of game mechanics, with which most gamers around the world are extremely familiar.

And thirdly, these phases are introduced as a type of hierarchy of how deep the content delves into local adaptation, these steps by no means happen sequentially. As we know, even localization takes place in various stages within the typical game development cycle, so culturalization is a coordination of various tasks and priorities being orchestrated across the entire development process. While we continue to strive to get game developers to perceive localization as a core and integral aspect to development, instead of something tacked on at the end, culturalization demands that international considerations start on Day One of the project. The most effective culturalization happens at the beginning, with a review and consideration of the game's overall design, direction, and theme, so that any potential concerns for a local market might be addressed. Typically, at least 75% of any potential issues can be easily managed in these early stages, which is advantageous considering that any changes late in the process become very costly and time consuming.

3. The Critical Context of Culture

The majority of individuals working in the game industry have a basic awareness of how cultural interaction works; in other words, they realize that an action in one context can lead to either positive and/or negative reactions in another context. Some reactions are predictable while others may appear to be completely irrational. The way in which a local gamer reacts has much to do with the cultural context in which they exist, built by various aspects including their religion, ethnicity, language, specific location, and so on. It's also important to consider the other contexts in which the gamer exists, such as their social connections, economic status, and level of education. The bottom line is that exist a lot of underlying reasons for why people in a specific culture react to game content in a certain way. It's critical to be mindful of this fact when considering a local market's reaction to a game; not everyone will react the same way or for the same reasons.

Culture can mean so many things. In this particular discussion, it's hopefully been clear that the definition follows the common anthropological concept that culture is the general aggregation of the beliefs, customs, and practices of a specific group of people. Such academic approaches are helpful as a starting point, but what does "culture" really mean in the context of producing video games or any digital content for that matter? First, let's consider these two basic definitions that are posed from the perspective of a creator of digital content:

Content: Information created for perpetuation and dissemination; in video games, it's anything a player will see, hear or read (essentially everything in the game, except for the underlying code).

Context: The circumstances or events that form a unique environment in space and time, within which information is created and managed.

When we then consider these two fundamental concepts from a geographical and sociological perspective, they can yield the following definition:

Culture: The accumulated, managed content of a specific context.

Granted that this may seem overly simplified, but the point is to consider "culture" from a game developer's perspective. And thus with definition of culture, we can actually look at any specific local culture as a combined set of "content assets" that clearly define the look, feel, sound, taste and general nature of the culture in very much the same way the content assets of a game define the look, feel, sound and general nature of the unique game world. Along with those assets come expectations for what will or will not fit within the norms of that culture. If we view culture in this way, it's often easier to perceive how the content assets of a game's designed world might conflict with the expectations for what fits in the content assets of a local culture. If the game contains a piece of content that doesn't fit with the culture's expectations or is noticeable enough to shock the gamer out of the game's context and reinforce their own cultural worldview, then a potential problem arises.

The positive news is that most experienced game players understand the difference between the game's context and their own culture's. Because they play regularly and they have likely played a lot of different types of games, they are much less likely to react negatively to a piece of content that might normally not fit within their expectations. For most gamers, the concern is usually focused on whether or not the gaming experience is fun – and less about if the content is potentially offending them.

However, the drawback is that most cultural backlash around video game content doesn't originate with gamers; rather it comes from the *unintended* audience surrounding those gamers. Most typically these are people who don't play games, who don't understand the content-context relationship between the game world and "real world," and who might often have a negative predisposition towards games. While the local gamer might see something offensive in a game and disregard it in favor of continuing the game, their parents, local lawmakers, clergy, and others may become outraged at what they see without taking the time to understand why it's in the game and what role it serves. Ironically, it's been observed that the greater the negative reactions of the unintended audience, the more interest gamers seem to take in the now-controversial game title. This is turn often encourages game developers to keep their content edgy, following the old notion that "there's no such thing as bad press." But over time, this pushing of cultural tolerances can have a detrimental effect for the game publisher in a specific market, such as creating distance with their intended gamer audience and possibly provoking government sanctions.

4. Types of Culturalization Challenges

In our emerging and globalizing information-based society, one essential realization is that *content carries culture*; it's a reflection of the culture in which it was created and it evokes reaction from the cultures to which it's distributed. Given this, it's often challenging for a game designer in one locale to be aware of and account for the issues that could cause problems in another locale. However, by considering the following broad categories of cultural aspects that most often generate conflict between the game's context and local cultures, it is possible to proactively reduce the potential for issues to arise.

4.1 History: Past and Present

Many cultures are extremely protective of their historical legacy and origins, thus any alternate or reimagined history can often yield strong, emotional reactions. Without a doubt, the issue of historical accuracy is one of the most sensitive issues for games with any basis in the actual world. Using the statistical concept of the "long tail", consider how many cultures celebrate major events in their past, even some that occurred many centuries ago. People long remember because such events helped to form the very nature of their particular culture and/or nationality and are integral to their identity.

Yet it's important to remember that the perception of history can often be clouded by the fog of the present day interpretation of what really happened decades or centuries ago, thus even an honest attempt to accurately portray a historical scenario in a game setting can go awry. For example, in *Age of Empires* (1997), a battle scenario was created in which the Yamato armies of Japan invaded the Korean peninsula and effectively overwhelmed the Chosen people of Korea. Historians tell us that this is essentially what happened and so the game designers diligently recreated the event. However, the government of South Korea saw history differently and disputed the scenario's accuracy as being historically inaccurate. The result was that a downloadable patch was created which changed the scenario so that instead of a Yamato invasion of Korea, it was a Chosen invasion of Japan, i.e., the Chosen armies were given a fighting chance in the game.

The impact of leveraging history in games has been seen in several titles, from other real-time strategy games like *Rise of Nations* (2003) and *Civilization* (1991) to combat-oriented titles like *Call of Duty* (2003) and *Medal of Honor* (1999). History is a very compelling topic for a game, however it's rarely possible to provide the full context of a historical event simply due to the scale and cost of game development. For the sake of game play, many details must be generalized which is one aspect that is often noticed – the lack of detail in some manner, from costumes to locations to the series of events. Employing history in a video game is much like a cartographer creating a map: the world must be generalized (i.e., remove information) in order for the map to be useful for its intended purpose.

Distant history isn't the only aspect that can be problematic. Recent history can also be a very sensitive topic as the events and outcome are very raw in people's minds. Depicting a recent conflict or otherwise traumatic event can often cause a swift backlash due to perceived insensitivity on the part of the game developer. In 2009, the title *Six Days in Fallujah* was a prime example of this. The title reenacted the late 2004 events of what was called Operation Phantom Fury by the U.S. military in which U.S., British and Iraqi troops attempted to remove the insurgents using Fallujah as their base. The fight proved to be costly for both sides and became surrounded with controversy over the alleged use of white phosphorus as a chemical weapon. The U.S. coalition forces suffered over 100 deaths and over 600 wounded, while estimates put the insurgent losses at over 1,300. If several decades had passed since this event, it no doubt might make a compelling game scenario but the recent nature of the event made it so potentially inflammatory that the publisher Konami decided not to publish the title in any market after it sensed growing anger over the use of this specific battle (and the title remains unpublished to this day).

4.2 Religion & Faith: Sacred vs. Secular

Game content creators need to be particularly sensitive to the underlying mechanics of the cultures into which their game titles are to be disseminated. If a specific culture has a more sacred basis for their daily activities and social administration, meaning a society based on religious principles and practices, then the rules pertaining to acceptability will be quite different from a culture based on a more secular outlook. In general, a society based on sacred rules tends to be less flexible and yielding to the context in which information appears because they are following what they consider to be a higher standard than human judgment; i.e., if problematic content appears *at all*, regardless of context, then there is potential for backlash.

As an example in contrast, China and the Middle East capture both the promise and fear of game content producers. The two regions are a study in contrast, not only from the obvious cultural differences and languages, but also from the underlying mechanisms that drive the respective societies. On one hand, the Chinese society is managed as a rather "secular" form of government with overt social engineering, while the Middle East (referring specifically to its predominant Islamic countries) is managed from a more "sacred" perspective – basing their government functions and society on the tenets of their religion. From a game developer's perspective, the quick judgment and/or rejection from either locality may seem very similar on the surface, yet the respective reactions are based on entirely different assumptions.

Clearly, as one might suspect, any piece of game content that challenges the tenets of a religion or a belief system is one of the most potentially hazardous issues. In 2002, work was progressing on a hand-to-hand fighting game named *Kakuto Chojin* that was envisioned as a great addition to the original Xbox game library. Everyone involved created a solid title, yet with one exception: a brief audio track was incorporated into the game that included a chanted portion of the Islamic Qur'an. While the error was fixed, some copies unfixed copies reached store shelves and the issue became widely known within a few weeks. *Kakuto Chojin* was subsequently banned in Saudi Arabia and a few other Muslim countries and the backlash became so widespread that the product was globally recalled and discontinued. Consider that this mis-use of the Qur'an is not too dissimilar from the extreme worldwide backlash from the Islamic community in 2005 that resulted from the editorial cartoons of Mohammad that published in a Danish newspaper. To followers of the specific faith, religious content is often a sign of hope and perseverance and generates deep feelings. Such highly-revered and widely recognized symbols do not allow much (if any) modification or reinterpretation; any such deviation is usually an invitation for severe backlash.

Even something as simple as religious architecture can be a problem. In 2006, *Resistance: Fall of Man* was released and the Church of England was shocked to learn that their Manchester Cathedral was recreated in great detail in the game's world. Not only was it recreated but some of the violent action of the game took place within the church itself, something considered very disturbing by the Church of England. Sony eventually issued an apology and the Church of England later issued new "Sacred Digital Guidelines" to help video game developers and others learn how to respectfully leverage their religious structures.

4.3 Ethnicity: Inclusion vs. Exclusion

Besides the more volatile issues of history and religion, there are a host of issues that are related to the portrayal of specific ethnic groups from around the world. Of course, there are also issues about the stereotyping of gender, age and so forth but ethnic stereotyping has been a particularly blatant issue.

Chief among these issues is the perceived inequitable treatment of a specific ethnicity and/or nationality. Typically, a specific group perceives that they are being included or excluded in the game content with some form of negative intent. Even before its release in early 2009, the title *Resident Evil 5* had generated significant negative publicity due to its perceived racism. In the game, the clean cut, white Caucasian protagonist is seen roaming through a village in sub-Saharan Africa and gunning down unarmed, obviously impoverished African villagers. Several commentaries on the imagery started to circulate and a broader discussion on the subject ensued in the press, in blogs and so forth. While the publisher Capcom was quick to point out that the African villagers were in fact infected, zombie-like humans who were out to kill the protagonist, the stark imagery of a white man killing black villagers evokes powerfully negative feelings from history. Notions of the "great white hunter", the "dark continent of Africa" and so forth quickly came to mind for many people, particularly the aforementioned unintended audience. The developers had a clear reasoning for the conflict within the game, which actually upon its release portrayed a greater diversity of zombie targets for the protagonist. But the backlash provides ample reason for a publisher to stop and question if mimicking that kind of negative imagery is appropriate, even if it's completely unintentional. Would portraying less inflammatory imagery have negatively affected sales in any way? It is highly doubtful because the intended gaming audience usually isn't going to complain about the content at this level, especially on such a highly anticipated title. The unintended, non-gaming unintended audience did take notice but the impact was negligible; the game was one of the most anticipated and financially successful titles of 2009.

Some other popular titles like the 2009 iPhone game *Pocket God* follow this notion of evoking antiquated stereotypes of specific ethnicities. In *Pocket God*, the player is essentially the "god" over a small fictitious island and has the ability to torment small indigenous people though activities such as feeding them to sharks, dropping them from great heights, getting a volcano to spew hot lava on them, having killer ants devour them, and so forth. When confronted about the portrayal of the natives, Bolt Creative, the game's developer, was clear that the designers did not intend to depict any specific nationality. However, with the various visual cues on the island, including a distinctive *moai* statue from Easter Island, the native outfits and their darker skin color, it was enough for Pacific Islander advocates to protest the game as a blatant continuance of the "primitive" ethnic stereotype.

Other forms of conflict over ethnicity and/or nationality might be rooted in history but are still problematic due to ongoing tension between two or more cultures. This was evident when *Age of Empires II: The Age of Kings* (1999) was released in South Korea. On the box art on the game's packaging was a depiction of three warriors from history: a Viking, an Anglo-Saxon and a Japanese Samurai. Due to the existing cultural friction between Korea and Japan at the time (related to heightened tensions related to territorial issues), going back deep into history including the period of Japanese aggression during World War II and several geopolitical incidents since, Korean retailers were reticent to place the game on their shelves because of the samurai image. When the *Age of Empires II: The Conquerors* (2000) expansion pack was released, the Korean box art prominently replaced the Aztec warrior that was seen in other locales with a

Korean general in front of two other historical soldiers, giving the packaging a definite Korean feel.

Unfortunately, the video game format is beginning to attract various dissonant groups who often possess less-than-mainstream views about ethnicity and nationality. The games medium is leveraged as a way to promote their specific cultural viewpoint and to help generate more appeal for their fringe cause. Many of these titles are easy targets for media outlets that often use them as examples of how the video game industry is inherently negative and debased. This isn't surprising when such games have included titles like *Muslim Massacre: The Game of Modern Religious Genocide* (2008), *Ethnic Cleansing: The Game* (2002; designed by and for neo-Nazis), *Special Force* (2003; created by the Hezbollah to train their youth on how to kill Israelis), and *Border Patrol* (2008; where the player tries to stop highly stereotyped Mexicans from crossing into the United States).

4.4 Geopolitics: Imaginations of Sovereignty

One of the most fundamental means by which national governments make themselves known to content developers of all varieties is by reinforcing their territorial sovereignty. This often involves geographic issues where the government claims certain territories and they expect those territories to be shown as integrated with their nation, whether it's on a functional map or in the world of a video game. For companies that are involved in online cartography, such as Google and Microsoft, there are persistent challenges from governments which seek to reinforce their perceived territorial control through these companies' maps – making requests for boundary changes, name changes, and so on. This local perception of sovereignty can be considered the "geopolitical imagination" of the government, because they actually do not control the areas they claim on their maps. With some governments, such as China and India, there is absolutely no room for negotiation on this issue; in fact they actually employ laws that dictate how their national maps must appear.

One key example of this kind of government enforcement occurred with the game *Hearts of Iron* (2002) and its sequel. This game's map was divided into somewhat random sectors in a fashion very similar to the classic board game *Risk* where the world is split into various chunks of territory, often not divided along existing country boundaries. In the case of this game which takes place during World War II, China was divided into several distinct pieces, including Tibet and Taiwan being shown as if they were separate countries. Even though the time context is World War II, China's government banned the game – even though neither Tibet nor Taiwan was part of the People's Republic of China at that time (the PRC didn't come into existence until 1949 and Taiwan presently remains apart from mainland China). The historical and geopolitical facts became secondary to the present government's need to reinforce their own perception of their territory in every possible context.

In slightly different type of geopolitical issue, the Korea Media Rating Board (KMRB) of South Korea was quick to ban *Ghost Recon 2* (2004) because the game featured a rogue North Korean general who had delusions of grandeur. The South Korean government has had a standing policy of opposing any representation of their northern neighbor as an aggressor because there are long-term hopes of an eventual reunification of the Korean peninsula; most Koreans view their country as north and south combined into one people – not divided as is currently the case. This may seem counterintuitive given the recent tensions between the two countries, but the goal remains intact. For this reason, the KMRB also banned the game *Mercenaries* (2005) and previously experienced public furor in 2002 with the James Bond film *Die Another Day* because the main antagonist was North Korean. Incidentally, in 2006 the Korean government creating the Game Rating Board (GRB) to specifically deal with video games and leave other types of content to the KMRB. After pressure from Korean gamers who were emphasizing their free speech rights, the GRB lifted the bans on *Ghost Recon 2* and *Mercenaries*.

History, religion, ethnic conflict and geopolitical friction – these truly are the top considerations when designing global-facing content for video games. Of course there are many subcategories and other related topics, but these are the most fundamental dimensions of game culturalization. Also, it's important to emphasize that it's almost never just a question of history or of geopolitics but rather a complex intertwining of all of these factors. With this plethora of potential sensitivities, it's likely easy to see why some game developers might choose to ignore such issues and simply hope for the best outcome. But as the saying goes, "an ounce of prevention is worth a pound of cure," thus adopting a proactive stance to such issues can place developers in a better position to develop games for the global audience.

5. Final Comments

The relatively small investment of time and effort of content culturalization during the game development process will offset a tremendous loss of time, money and public relations in having to resolve a post-release cultural issue in a game. How might game designers incorporate culturalization practices into their work? The limits of this space prevent a thorough look at the various types of culturalization methods, which generally involve the following steps:

1. *Awareness:* A major achievement is the simple step of attaining a basic awareness of the potential for cultural issues. Content creators and managers need to have at least a rudimentary understanding that cultural issues can occur, and then even better if they know what specifically might be a problem in which local markets.
2. *Identification:* At this stage it's important to apply the notion of *contextual proximity* as a guiding method. Stated simply, contextual proximity is the concept that the closer a content element approaches the original context in person, place, time and/or form, the greater the potential for cultural sensitivity.
3. *Assessment:* At the most basic level of triage, it's helpful to separate "overt offenses" – the obvious things that you know for certain will be a problem from the "reasonable risks" – the things that might raise some concerns but won't likely prevent a game from remaining in the intended market.
4. *Implementation:* Employ one of the most important principles of culturalization: *make the most minimal changes to the least amount of content.* Only change what really must be changed in order to ensure distribution to the game's target market, while also being mindful of the effects on surrounding locales.

Lastly, it's imperative to stress that whatever geopolitical and cultural issues may be discovered in a game, every game publisher has a choice as to whether or not to make any changes. The majority of companies choose to do so because it means preserving a key market or opening up a new one for business. However, there are times when it may not make sense to make even a small content change. In cases such as this, it becomes absolutely critical to document all decision-making in the form of a solid defensive explanation. This shouldn't be a long treatise on how the content choices were made, but rather a straightforward rationale which explains why certain content choices were made.

As is often common in the U.S. and western companies, leveraging the "freedom of speech" argument won't go very far with many local governments because in many locales around the world that concept carries little meaning. Instead, focus on the fact that careful research was conducted and a decision was reached that serves the company's goals and the local consumers' expectations. This form of documentation may not save a company from all repercussions, but showing that content decisions were conscientious rather than random can aid in diminishing government and consumer backlash.

Well-executed culturalization within a game development cycle is not something that happens overnight, nor without some commitment of some resources, but it can prove to be an invaluable long-term return on investment. The benefits to a company's content quality, government relations, and public image amongst local consumers will become evident over time. Game developers should create the game they wish to create, but without forgetting the multicultural audience who will be participating in that creative vision. While culturalization may seem like a sure way to stifle the creative forces behind game development, it's actually a path to ensuring that the enjoyment of that creative vision can be maximized to as many cultures as possible. In the end, the real key to the culturalization method is to simply respond respectfully and proactively to the local market's perception of the developer's intentions, i. e. just take even a brief moment to view the game content from their eyes.

Notes on Contributors

VÍCTOR ALONSO LION has been dedicated to internationalization and localization for more than a decade. His experience covers, among other things, localization engineering, international project management, localization training and globalization consultancy. He provides project management to a wide range of areas, such as life sciences, IT, the financial industry, and of course to the e-learning, entertainment and game industry. Víctor is a frequent speaker on international project management and localization.

EUGENIA ARRÉS is a technical translator and localizer and she has been working as freelance translator and proofreader since 2003. Besides, she has worked as an in-house localization project manager for 5 years. Her academic training includes translation, interpreting and applied languages (she holds 3 European Bachelors) and she is currently following her PhD courses in Translation and Interpreting Research at Universidad de Granada; she is also a TILP member and achieved the Certified Localization Professional degree in 2010. She has been training translation students at graduate and postgraduate level since 2007 at Universidad de Granada and Universidad Pablo de Olavide, Sevilla. She won in 2012 the award to the best "Teaching Innovation Practices" given by Universidad Europea de Madrid. She has authored or co-authored several papers on various aspects of Applied ICTs and Translation Didactics.

OLIVER CARREIRA (http://olivercarreira.net) is an EN-ES Localization & Translation Professional, working in the language industry since 2007. His main areas of expertise include software, website and mobile app localization, technical and life sciences translation, marketing and transcreation. He also works as a teacher, mainly offering training oriented to professional translators. As a researcher, he has published a review on the crowdsourced translation process applied by Facebook to the contents of its platform (http://dialnet.unirioja.es/servlet/articulo?codigo=4181841). His main research interests include localization and translators training. He is also member of the publishing team of *La Linterna del Traductor*, the biannual magazine published by ASETRAD (Spanish Association of Translators, Interpreters and Proofeaders).

ÁNGEL DEL BLANCO finished his Master Degree in Computer Science at the UCM in 2009, and currently he is working as a full-time researcher in the e-UCM research group as well as being a PhD student. His research focuses on the technical integration of highly interactive contents into e-learning environments, with special emphasis in the current e-learning standards and its limitations. He has published 28 research papers in academic journals and conferences in the field.

KATE EDWARDS is the Executive Director of the International Game Developers Association (IGDA). She is also the founder and principal consultant of Geogrify, a Seattle-based consultancy for content culturalization, and a unique hybrid of an applied geographer, writer, and corporate strategist, built upon a passion for global cultures and media technologies. Formerly as Microsoft's first Geopolitical Strategist in the Geopolitical Strategy team she created and managed, Kate was responsible for protecting against political and cultural content risks across all MS products and locales. Since leaving Microsoft, she has provided guidance to many companies on a wide range of geopolitical and cultural issues, and she continues to work on games such as the Dragon Age series, Modern Warfare 3, Star Wars: The Old Republic, Dance Central series, Mass Effect 3 and Halo 4. Kate is also the founder and former chair of the IGDA's Game Localization Special Interest Group, the co-organizer of the Game Localization Summit at GDC, and is a regular columnist for MultiLingual Computing magazine.

ALBERTO FERNÁNDEZ COSTALES graduated in English Philology from the University of Oviedo where he obtained his PhD with a thesis on the internationalisation and localisation of university websites. He has been a visiting scholar at Imperial College London and since 2011 he is a member of the Transmedia Catalonia Research group. His research interests include game localisation and accessibility, audiovisual translation, language technology, language teaching, bilingualism and multilingualism. He is a lecturer and researcher at the University of Oviedo where he teaches applied linguistics.

BALTASAR FERNÁNDEZ-MANJÓN is a doctor and a full professor at the Complutense University of Madrid. His main research interests are e-learning technologies, educational uses of serious games (e.g. in the medical domain), application of educational standards. He leads the e-UCM research group where the eAdventure environment for developing educational games and simulations has been developed. Dr Fernandez-Manjón is IEEE senior member and associate editor of IEEE Transaction on Learning Technologies. He is currently involved in several EU projects about serious games: GALA NoE and LLP SEGAN and CHERMUG.

DIMITRIS GRAMMENOS is a Principal Researcher at the Institute of Computer Science (ICS) of the Foundation for Research and Technology – Hellas (FORTH). He is the lead interaction designer of the Human-Computer Interaction (HCI) Laboratory, specializing in the domains of Universal Design, Universal Access and Ambient Intelligence Environments. He holds a B.Sc in Computer Science (Ptychion, University of Crete, Greece), and an M.Sc and Ph.D in Electronic Engineering (Department of Electronic Engineering, University of Kent, UK). From 2006 to 2010, he has been in charge of the HCI Laboratory's Universally Accessible Games Activity, actively pursuing the application and integration of Design for All principles in the field of computer games through the development of both theoretical research work and related applied examples. Currently, as a member of the Ambient Intelligence (AmI) Programme of ICS-FORTH, he is heavily involved in the design and development of novel, human-centred, AmI environments and interactive applications.

ORNELLA LEPRE is a PhD candidate in Translation Studies at Imperial College London, where her doctoral thesis deals with the translation of humor in video games. Her main research interests are audiovisual translation, with particular attention to video game localization, and the cultural implications of translation. She is also interested in the application of methods first developed in Economic Theory and Statistics Linguistics and Translation; she holds an MSc and a PhD in Economics from the University of Naples "Federico II" and an MRes degree from the London School of Economics and Political Science. She has presented at several international conferences and taught in various universities in Italy and the UK. She is also a freelance translator, specializing in video game and software localization.

JAVIER MAIRENA is a game programmer that has specialized in video games accessibility, prototyping, documenting, and participating in several conferences on accessibility and electronic entertainment; as well as writing on the game accessibility Castilian blog <www.videojuegosaccesibles.es>. He currently works in AccessAble Games (www.accessablegames.com) and The Game Kitchen (www.thegamekitchen.com) as responsible and accessibility expert in video games.

STEPHEN MANDIBERG is currently finishing a PhD in Communication at the University of California, San Diego. His research is about how the video game industry treats translation like a commercially neutral practice instead of a culturally productive one, and how games are understood not as an artistic medium of communication that can shape cultural bodies, but as commercial toys that are put on shelves for sale. His work more broadly is about the translation, transference and repetition of cultural texts (including new media, cinema and gaming) between distinct, national/cultural entities and over time.

CARMEN MANGIRON completed her PhD in Translation Studies at the Universitat Autònoma de Barcelona (UAB). After working in Dublin City University for a number of years, she currently works as a lecturer and researcher at the Universitat Autònoma de Barcelona. She is a member of the research group TransMedia Catalonia, part of the Centre for Ambient Intelligence and Accessibility in Catalunya. She has participated in several international conferences and has extensive experience as a translator, specialising in software and game localisation. Her main research interests are game localisation, game accessibility, and serious games. She is co-author of *Game Localization: Translating for the Global Digital Entertainment Industry* (O'Hagan and Mangiron, 2013).

IVÁN MARTÍNEZ-ORTIZ is an Associate Profesor in the Computer Science Studies at UCM and a member of the e-UCM research group. His research interests include e-learning technologies and the integration of educational modeling languages, serious games and e-learning standardization. Within the group, he has led the development of tools to support diverse IMS educational specifications including the integration of games in the learning flow. He has contributed more than 30 academic publications in topics related the Technology Enhanced Learning domain.

PABLO MORENO is a doctor and an Associate Professor in the Department of Software Engineering at Complutense University of Madrid. His research interests focus on educational uses of video games and game-like simulations. He has published over 50 academic articles on topics covering e-learning, integration of games in e-learning environments, and using games as learning evaluation tools. He is currently involved in several EU projects about serious games: GALA NoE and LLP SEGAN and CHERMUG.

RAFAEL MÜLLER GALHARDI is an English into Brazilian Portuguese translator and holds a master in Translation and Accessibility. He has recently defended his master's dissertation about *Chrono Trigger*'s fan translations in Brazil and has been working with game translation and localization for the past two years. He first started as an undergraduate student interested in analyzing fan translations, then took part and presented his projects at two editions of Fun for all: International Conference on Video Game and Virtual Worlds Translation and Accessibility in Spain and also at other scientific events in Brazil. Meanwhile, he also has become involved in many translation projects ranging from Facebook games to Multiplatform ones.

MINAKO O'HAGAN, PhD, is a Senior Lecturer in the School of Applied Language and Intercultural Studies (SALIS), Dublin City University. Since joining SALIS in 2002, she has expanded her main field of research Translation Technology into Video Game Localisation. Her PhD supervision covers a range of emerging areas of research, including an eye-tracking study of abusive subtitling of Japanese anime and a netnography-based study of Facebook translation crowdsourcing. Her recent publications include Linguistica Antverpiensia New Series – Themes in Translation Studies: Translation as a Social Activity – community translation 2.0 (O'Hagan [ed.] 2011) and a co-authored monograph *Game Localization* (O'Hagan and Mangiron 2013).

PILAR ORERO PhD (UMIST) is the head of research at CAIAC Research Centre (Universitat Autònoma de Barcelona, Spain). Director of the European MA in Audiovisual Translation at UAB, <http://mem.uab.es/metav/>. Recent publications: *Topics in Audiovisual Translation* (2004) John Benjamins. Co-editor with Jorge Díaz-Cintas and Aline Remael of *Media for All: Subtitling for the Deaf, Audio Description and Sign Language* (2007) Rodopi. Co-editor with Anna Matamala of *Listening to Subtitles: SDHoH* (2010) in Peter Lang. Co-writer with Anna Matamala and Eliana Franco of the *Voice-over: An Overview* (2010) in Peter Lang. Co-guest editor with J. L. Kruger Perspectives on *Audio Description* (2010). Leader of numerous research projects funded by the Spanish and Catalan Gov. Leads TransMedia Catalonia <http://grupsderecerca.uab.cat/transmediacatalonia> and is now part of the ITU working group Audiovisual Media Accessibility <http://www.itu.int/en/ITU-T/focusgroups/ava/Pages/default.aspx>. She holds the INDRA accessibility chair 2013.

GIANNA TARQUINI is a Postdoctoral Research Fellow in the Department of Interdisciplinary Studies in Translation, Languages and Culture (SITLeC) of the University of Bologna, Forlì campus. She has developed a research project on video game localisation, partly investigated in Japan, Kwansei Gakuin University, for the Game Studies area, and at the Localisation Research Centre of the University of Limerick, for the GILT modus operandi. She has a four-year experience in the industry as a linguistic localisation tester and freelance translator. Her research interests also include linguistics and audiovisual translation, which she is currently investigating in a postdoc project on the Forlixt multimedia database for AVT research.

JAVIER TORRENTE works as a contract researcher for the University Complutense of Madrid, in the group of educational technologies and e-learning e-UCM directed by Dr. Baltasar Fernández-Manjón. His research mainly focuses in the field of Game-Based Learning in general and the eAdventure project in particular. His current research includes universal design of digital games, personalized learning, mobile learning and e-Learning standards. He's coauthor of more than 50 academic papers published in international conferences and journals of the field.

ANNELIES VAN OERS, Annelies van Oers, MA has been a translator of videogames from English to Dutch for the past seven years at the Dutch videogame localisation company U-TRAX M.M.L. and has translated many AAA video games to Dutch. After her BA in English at the University of Utrecht, she earned her Master's degree at the Vrije Universiteit of Amsterdam. In her master's thesis she analysed the Dutch localisation of the video game *Beyond Good and Evil* (Ubisoft, 2003) by means of several translation strategies. She regularly gives guest lectures on the subject of video game localisation and continues to work as a video game translator.

XIAOCHUN ZHANG is currently working as a research assistant while doing a PhD at the University of Vienna, Austria. She was educated at Nanjing Normal University (BA in Applied English), P. R. China, and received her MA in Translation Studies from the University of Edinburgh, United Kingdom. Her research interests lie primarily in audiovisual translation with a specific interest in video games localization. Other areas of interest include language technology and terminology management.

www.ingramcontent.com/pod-product-compliance
Ingram Content Group UK Ltd.
Pitfield, Milton Keynes, MK11 3LW, UK
UKHW021845100926
8556IPUK00019B/1063

9 783034 31450